Advances in Geographical and Environmental Sciences

Advances in Geographical and Environmental Sciences synthesizes series diagnostigation and prognostication of earth environment, incorporating challenging interactive areas within ecological envelope of geosphere, biosphere, hydrosphere, atmosphere and cryosphere. It deals with land use land cover change (LUCC), urbanization, energy flux, land-ocean fluxes, climate, food security, ecohydrology, biodiversity, natural hazards and disasters, human health and their mutual interaction and feedback mechanism in order to contribute towards sustainable future. The geosciences methods range from traditional field techniques and conventional data collection, use of remote sensing and geographical information system, computer aided technique to advance geostatistical and dynamic modeling.

The series integrate past, present and future of geospheric attributes incorporating biophysical and human dimensions in spatio-temporal perspectives. The geosciences, encompassing land-ocean-atmosphere interaction is considered as a vital component in the context of environmental issues, especially in observation and prediction of air and water pollution, global warming and urban heat islands. It is important to communicate the advances in geosciences to increase resilience of society through capacity building for mitigating the impact of natural hazards and disasters. Sustainability of human society depends strongly on the earth environment, and thus the development of geosciences is critical for a better understanding of our living environment, and its sustainable development.

Geoscience also has the responsibility to not confine itself to addressing current problems but it is also developing a framework to address future issues. In order to build a 'Future Earth Model' for understanding and predicting the functioning of the whole climatic system, collaboration of experts in the traditional earth disciplines as well as in ecology, information technology, instrumentation and complex system is essential, through initiatives from human geoscientists. Thus human geoscience is emerging as key policy science for contributing towards sustainability/survivality science together with future earth initiative.

Advances in Geographical and Environmental Sciences series publishes books that contain novel approaches in tackling issues of human geoscience in its broadest sense—books in the series should focus on true progress in a particular area or region. The series includes monographs and edited volumes without any limitations in the page numbers.

Ashoka G. Dessai

Environment, Resources and Sustainable Tourism

Goa as a Case Study

Springer

Ashoka G. Dessai
School of Earth, Ocean and Atmospheric Sciences
Goa University
Taleigao Plateau, Goa, India

ISSN 2198-3542 ISSN 2198-3550 (electronic)
Advances in Geographical and Environmental Sciences
ISBN 978-981-99-1842-3 ISBN 978-981-99-1843-0 (eBook)
https://doi.org/10.1007/978-981-99-1843-0

This Springer imprint is published by the registered company Springer Nature Singapore Pte Ltd.
The registered company address is: 152 Beach Road, #21-01/04 Gateway East, Singapore 189721, Singapore

To
My Grand Aunt—Āka (Jaya)
and
Grand Daughter—Bālu (Arinya)

Preface

Tourism is one of the fastest growing sectors of the global economy. It is viewed as an important source of income, employment and an excellent revenue earner particularly by the developing economies. However, its rapid expansion, although considered a possibility for sustainable development, imposes tremendous stress on the environment, energy and other natural resources apart from impacting the local culture.

The world economy largely depends on mineral materials. The supply of minerals to the industry has become a global concern in the twenty-first century. Large demography, rapid pace of development and excessive consumption of resources in the developed and developing economies pose severe threats to mineral resources and environment. Earth resources are finite and substitutes for important minerals are scarce. Many minerals such as fossil fuels cannot be recycled. Therefore, improvement in resource efficiency is a prime requisite which will enable us to move towards the 1.5°C goal by contributing to climate-change mitigation.

This book lays special emphasis on the Protection and Preservation of the Environment, Conservation of Resources and at the same time delves into the environmental and industrial factors that govern the production and optimal consumption of mineral resources of Goa. It discusses the characteristics of its mineral resources and gauges the impact of relentless exploitation of minerals through rapid industrialization for over six decades. It deals with overexploitation and consequent pollution of the water resources already subjected to degradation from industrial effluents. The impact of all these issues on the physical environment of the State is a matter of great concern. The book examines the present tourism situation and the potentialities of the State to explore means of conservation of the resources by recycling and reusing thereby extending the life of the resource base of the State. Based on the analysis of the resource potential, the rate of utilization and the consumption scenario in the State, a set of recommendations is proposed that could be taken into consideration by the State and the International (Donor Institutions) stakeholders for promoting tourism as a tool for sustainable development. Finally, the effect of excessive tourism on the depletion of natural resources and its impact on the socio-cultural environment and life styles of the local populace are outlined. This information is essential, if we are to understand all the facets of the issue at hand and hopefully, find a solution in the interest of all concerned.

The book evolves from the presentation of an updated account of the mineral resources of the State. A concise summary of the major resources of iron, manganese, bauxite, soils, sand and marine shells is provided to evaluate their potential for economic development.

Goa, as a State subjected to heavy mining activity for over seven decades, has been left with deep scars on the natural environment in general, and the social environment in particular. The chapters on mining industry and environmental assessment attempt to interconnect the exploitation of mineral deposits and its impact on the environment. The contents of this chapter were, however, collated when the mining activity was at its peak, that is, before the imposition of ban on mining (September 2012) by the Hon. Supreme Court of India. Hence, the book depicts the mineral exploitation situations, the facts and figures as prevailing then.

Water is the single commodity which controls the sustainability of the economy of a region, nation or the world. The chapter on water resources of the State discusses the potential of the resource, the rate of consumption, the spacio-temporal variation, the causes of depletion, the measures of conservation and revitalization of the resource. The impact of pollution on the quality of the resource and the necessary mitigation measures to arrest further deterioration and protection from degradation are outlined.

The chapter on environmental assessment critically examines the various types of stresses on the environment in general, and on the earth resources in particular, as a result of proliferation of anthropogenic activity. It not only seeks to suggest ways and means of conserving the environmental resources and protecting the ecology, but also discusses the implications of excessively rapid pace of resource depletion and its effects on the physical, socio-economic environment and the moral fabric of the State. Increased solid waste, excessive pollution and augmented emissions are the other threats to the environment. Various types of waste materials and their disposal have been distressing issues of the State since decades. The different aspects of waste disposal, industrial pollution and sustainable waste management are discussed in the chapter on waste management.

Tourism industry is the major revenue earner of the State. It has also been a formidable consumer of the State's natural resources at quite a rapid pace. Since the impact is on resources, it has an influence on the physical-, socio-economic and socio-cultural environment of the State. The last chapter critically analyses the tourism sector in the State, and how it could be made a sustainable activity to foster sustainable development of this biodiversity hotspot by providing in-depth view on the peoples of different heus and their role in safeguarding the future of this unique region.

The existing information on the natural resources and environment is sparse, scattered, quite often significant but rather passé. A usable, pertinent book which could provide a comprehensive discourse integrating environmental and socio-economic perspective on the complex issue of sustainable tourism and be a useful source of reference to the students, academics, professionals, planners, policy-makers and the general readership, was felt necessary. An attempt is also made to update, collate, synthesize the available information with authorial insight and interpretation, and present it in a logical manner, easy for the benefit of even a novice in earth science, environmental science and social sciences and humanities.

Taleigao Plateau, India Ashoka G. Dessai

Acknowledgements

The manuscript of the book is greatly benefitted from insightful reviews and valuable suggestions from Albertina Almeida, S. M. Borges, B. Krishna Rao and T. M. Mahadevan. I am extremely grateful to them for their useful comments and for sparing precious time.

I would like to thank D. B. Arolkar, B. R. Jayanth and M. G. Kale, my former students, colleagues and friends who have helped me in various ways during the preparation of the manuscript.

I have drawn freely from published works, yet the references in the text are kept to the minimum in order to maintain the flow of the narrative. However, a select bibliography, fairly adequate to refer to recent works and new data, is provided at the end of each chapter.

I must mention here that my friend and colleague, Anthony Viegas, has rendered invaluable help in various ways during the preparation of the manuscript. But for his selfless help, it would have been impossible to complete this manuscript in time.

Ashoka G. Dessai

Contents

1 Introduction 1
1.1 Preamble 1
1.2 Background of the Mineral Resources 5
1.3 Physiography 6
1.3.1 Western Ghats Escarpment 6
1.4 Drainage 7
1.4.1 Tidal Creeks 8
1.4.2 Linearity of the West Coast 9
1.4.3 Beaches 10
1.4.4 Lakes, Tanks and Reservoirs 10
1.5 Climate and Vegetation 11
References 12

2 Mineral Resources 15
2.1 Introduction 16
2.2 Western Dharwar Craton 16
2.3 Shimoga-Goa Schist Belt 18
2.4 Classification of the Greenstones 18
2.5 Revised Stratigraphic Framework 19
2.6 Palaeoenvironments 19
2.7 Mafic Intrusive Rocks 22
2.8 Iron Ores 23
2.8.1 Ore Deposits: Geological Framework 24
2.8.2 Structural Disposition 24
2.8.3 Reserves 26
2.8.4 Production and Exports 27
2.9 Manganese Ores 27
2.9.1 Manganese Ore Resources 28
2.9.2 Reserves 29
2.10 Laterites and Bauxites 30
2.10.1 Bauxite Resources 30

2.10.2 Reserves ... 32
2.11 Soil Resources of the State ... 32
2.11.1 Distribution and Characteristics ... 33
2.12 Sand and Gravel Resources ... 33
2.13 Fossil Marine Shell Resources ... 35
2.13.1 Resource Characteristics ... 36
References ... 37

3 Mining Industry ... 39
3.1 Introduction ... 40
3.2 Environmental Geoscience ... 40
3.3 Concept of Systems ... 41
3.3.1 Types of Systems ... 43
3.4 Relationship Between Natural Resource and Environment ... 44
3.5 Global Efforts at Protection of the Environment ... 44
3.6 Modern Concepts and Ideas ... 45
3.6.1 Sustainable Development ... 46
3.6.2 Sustainability versus Growth ... 48
3.7 Environment vis-à-vis Mining: Indian Context ... 50
3.7.1 Measures at Protection of the Environment ... 51
3.7.2 Mineral Resources ... 51
3.7.3 Mineral Conservation ... 51
3.7.4 The National Mineral Policy ... 53
3.7.5 Mines and Minerals Regulation and Development Act ... 56
3.8 Mining of Mineral Resources ... 58
3.8.1 Mining Activity in Goa ... 58
3.8.2 Relationship of Type of Mining and Extent of Damage ... 60
3.8.3 Slope Stability ... 62
3.8.4 Factors Responsible for Slope Movement ... 63
3.8.5 Role of Water in Mass Movement ... 63
3.8.6 Slope Failure ... 64
3.8.7 Impact of Slope Failure ... 65
3.8.8 Type of Mass Movement ... 65
3.8.9 Management of Slope Stability ... 66
3.8.10 Preventive Measures ... 67
3.9 Opencast Mining ... 68
3.9.1 Limitations of Open Cast Mining ... 69
References ... 69

4 Water Resources ... 71
4.1 Introduction ... 72
4.2 Climate and Weather Conditions ... 72
4.3 The Science of Water ... 73
4.3.1 Concepts in Hydrology ... 74

4.3.2 Properties of Aquifer ... 76
4.4 Surface Water Resources ... 78
4.4.1 Surface Water Resources of Goa ... 79
4.5 Groundwater Resources ... 80
4.5.1 State Groundwater Resources ... 81
4.5.2 District-wise Groundwater Distribution ... 83
4.5.3 Spatiotemporal Variation of Groundwater ... 84
4.5.4 Sustainability vs Resource Depletion ... 87
4.5.5 Mine Dewatering: Implications on the Watertable ... 93
4.6 Issues Related to the Salinity of Water ... 96
4.6.1 Saltwater Intrusion ... 97
4.6.2 Groundwater-Related Salinity Issues ... 97
4.6.3 Tidal Impact ... 100
4.6.4 Factors Governing Seawater Intrusion in the State ... 100
4.7 Confronting the Water Challenge: Global Scenario ... 101
4.7.1 Integrated Water Resource Management ... 102
4.8 Sustainable Watershed Development ... 104
4.8.1 Water Conservation in the State: Background ... 104
4.8.2 Post-liberation Scenario: Interlinking of Rivers ... 105
4.9 Groundwater Conservation Initiatives ... 107
4.9.1 Conjunctive Use of Water ... 108
4.9.2 Artificial Recharge ... 109
4.9.3 Rainwater Harvesting ... 110
4.9.4 Mitigation of Water Crisis ... 111
References ... 112

5 Environmental Assessment ... 115
5.1 Introduction ... 116
5.2 Environmental Concerns ... 116
5.2.1 Control on the Components of the Environment ... 117
5.2.2 Societal Change Leading to Environmental Issues ... 118
5.2.3 Environmental Challenges ... 118
5.2.4 Threat to Aquatic Environments ... 123
5.3 Impact on Groundwater Quality ... 132
5.3.1 Quality of State Groundwater ... 133
5.3.2 Contamination of Groundwaters ... 134
5.3.3 Saltwater Intrusion: Need for Efficient Management ... 136
5.3.4 Availability versus Demand: Changing Sustainability ... 138
5.3.5 Quality of Surface Water ... 140
5.3.6 Degradation of Coastal Aquifers ... 141
5.3.7 Vulnerability Assessment ... 142
5.3.8 Mitigation Measures ... 143
5.4 Air Pollution ... 143
5.4.1 Threat to Human Health ... 147

5.5 Environmental Management ... 149
5.5.1 Cost–Benefit Analysis ... 150
5.5.2 Rehabilitation of Mined Areas ... 153
5.5.3 Restoration Initiatives ... 154
5.5.4 Sustainable Mining and Protection of the Environment ... 154
5.5.5 Underground Mining and Excavations ... 156
5.6 Construction and Industrial Mineral Resources ... 157
5.6.1 Sand Extraction: Environmental Challenges ... 158
5.6.2 Policy Framework ... 159
5.6.3 Guidelines ... 160
5.6.4 Management Plan ... 161
5.7 Fossil Marine Shell Extraction: Environmental Concerns ... 161
5.7.1 Geological Perspective ... 162
5.7.2 Habitat ... 163
5.7.3 Employment Opportunities ... 163
5.7.4 Statutory Framework ... 164
5.7.5 Environmental Strategies ... 164
5.7.6 Environment Management Plan ... 164
5.8 Concerns of Rising Sea Level ... 166
5.8.1 Vulnerability Assessment ... 166
References ... 167

6 Waste Management ... 171
6.1 Introduction ... 171
6.2 Municipal Solid Waste: Disposal Concerns ... 172
6.2.1 Solid Waste Management Policy Framework ... 172
6.2.2 Solid Waste Management of the State ... 173
6.2.3 Pollution from Leachates ... 173
6.2.4 Products of Waste Incineration ... 174
6.2.5 Biodegradable Waste ... 175
6.2.6 Biomedical Waste ... 175
6.2.7 Sustainable Management of CDW ... 176
6.2.8 Pozzolans in Light Weight Concrete ... 178
6.2.9 Electronic Waste ... 179
6.2.10 Plastic Waste ... 180
6.2.11 Glass Waste ... 181
6.2.12 Tyre-Rubber Waste ... 182
6.3 Industrial Pollution: Concern of Effluents ... 182
6.3.1 Type of Effluents ... 183
6.4 Overstressed Land Resources ... 184
6.4.1 Growth Pattern ... 184
6.4.2 Density of Structures ... 185
6.4.3 Unaffordable Land Prices ... 185
References ... 186

7 Sustainable Tourism ... 187
7.1 Introduction ... 188
7.2 Overview of Tourism Industry ... 188
7.2.1 Definition of Sustainable Tourism ... 189
7.2.2 Sustainable Development Index ... 191
7.2.3 State Tourism Policy ... 192
7.2.4 Principles and Objectives of the State Policy ... 193
7.3 Tourism: The Case of Goa ... 194
7.3.1 Demographic Pressures on Resources ... 195
7.3.2 Sustainable Tourism: Potential for State Initiatives ... 196
7.3.3 Coastal Resource System and Tourism Products ... 198
7.4 Impact of Tourism: Stressed Environment ... 200
7.4.1 Distress on Land Resources and Mangroves ... 200
7.5 Congested Road Traffic: Remedial Measures ... 203
7.5.1 Increased Emissions ... 205
7.5.2 Growth in Road Accidents ... 205
7.5.3 Challenges to Change ... 205
7.5.4 Improvement of Infrastructure ... 206
7.6 Construction Industry ... 206
7.6.1 Real Estate and the Workforce ... 207
7.6.2 Construction Materials and Substitutes ... 207
7.7 Socio-Cultural Conflicts ... 208
7.8 Socio-Economic Constraints ... 209
7.8.1 Constraints in Sharing Resource ... 209
7.8.2 Friction Between Communities ... 209
7.9 Education vis-à-vis Employment ... 210
7.10 Shifting Occupation Pattern ... 211
7.10.1 Fisheries Sector ... 211
7.10.2 Outmigration of Locals ... 212
7.10.3 Unviable Agriculture ... 213
7.11 Changing Lifestyle ... 213
7.11.1 Migration to Cities ... 214
7.12 Concerns of Fragile Economy ... 214
7.13 Increase in Illegal Land Conversions and Land Disputes ... 215
7.13.1 Policy Amendments ... 216
7.14 Challenges to Development Initiatives ... 216
7.14.1 Conflicts with the Industry ... 217
7.14.2 Waste Disposal ... 217
7.14.3 Public Infrastructure ... 218
7.14.4 Resource Allocation ... 219
7.15 Future Perspective and Recommendations ... 221
References ... 226

Index ... 229

Abbreviations

BHQ	Banded Haematite Quartzite
BIF	Banded Iron Formations
BMQ	Banded Magnetite Quartzite
BMUC	Bondla Mafic-Ultramafic Complex
CDW	Construction and Demolition Waste
CITZ	Central Indian Tectonic Zone
CPTPP	Progressive Agreement for Trans Pacific Partnership
CRZ	Coastal Regulatory Zone
DDD	Dichlorodiphenyldichloroethane
DDE	Dichlorodiphenyldichloroethylene
DDT	Dichlorodiphenyltrichloroethane
DSS	Deep Seismic Soundings
EDC	Eastern Dharwar Craton
EPA	Environment Protection Agency
FTAA	Free Trade Area of the Americas
HTL	High Tide Line
IPEF	Indo Pacific Economic Framework
LED	Light Emitting Diode
MP	Microplastic
MSW	Municipal Solid Waste
NIO	National Institute of Oceanography
NP	Nonylphenol
NPt	Nanoplastic
PCB	Polychlorinated Biphenyls
PET	Polyethylene Terephthalate
POP	Persistent Organic Pollutant
PP	Polypropylene
RCC	Reinforced Cement Concrete
RCEP	Regional Comprehensive Economic Partnership
SDI	Sustainability Development Index
SEZ	Special Economic Zone

SMM	Sustainable Material Management
UNWTO	United Nations World Tourism Organization
WDC	Western Dharwar Craton
WHC	World Heritage Convention
WTO	World Trade Organisation
WTTC	World Travel and Tourism Council

List of Figures

Fig. 1.1 Location map of Goa 3
Fig. 2.1 Geological sketch map of Goa (modified after Dessai 2018) 17
Fig. 2.2 Generalized lithosection for the supracrustals from Goa (after Dessai 2018) 21
Fig. 2.3 Banded Haematite Quartzite (BHQ), Bicholim Formation, (Ponda Group), Bicholim, Goa, consisting of alternate layers of brown haematite and white cherty silica (after Dessai 2018) 25
Fig. 2.4 Sand loading activity at a jetty from the Cumbarjua Canal in the vicinity of Amona 35
Fig. 3.1 Relationship of the perpetual, renewable and non-renewable resources with time. Perpetual resource can increase continuously with time. The non-renewable resource tapers with time. Renewable resource can continue provided the rate of production is identical to the rate of renewal (modified after Merritts et al. 1998) 47
Fig. 3.2 Large, extensive dumps of overburden in North Goa 60
Fig. 3.3 An open cast iron ore mine in North Goa depicting contour strip mining 61
Fig. 4.1 The aquifer types and the groundwater-seawater interface (modified from Wikipedia.org) 75
Fig. 4.2 Distribution of water on Earth (data source Shiklomanov [1993]; modified after United States Geological Survey [2019]) 79
Fig. 4.3 Map depicting the watersheds of the major rivers of the State (after Water Resources Department, Government of Goa) 80
Fig. 4.4 Pre-monsoon (May 2016) depth to water level map of aquifer I (modified after CGWB 2019) 86
Fig. 4.5 Pre-monsoon (May 2016) depth to water level map of aquifer II (modified after CGWB 2019) 87

Fig. 4.6 Post-monsoon (November 2016) depth to water level map of aquifer I (modified after CGWB 2019) 88
Fig. 4.7 Post-monsoon (November 2016) depth to water level map of aquifer II (modified after CGWB 2019) 89
Fig. 4.8 Decadal mean water level fluctuation (May 2006–May 2015) map of aquifer I (modified after CGWB 2019) 90
Fig. 4.9 Decadal mean water level fluctuation (Nov. 2006–Nov. 2015) map of aquifer I (modified after CGWB 2019) 91
Fig. 4.10 Map depicting the estuarine stretches and adjoining salinity-affected areas of the various watersheds from the State (modified after CGWB 2013) 98
Fig. 4.11 A sketch to illustrate seawater intrusion in coastal aquifer (modified after Prusty and Farooq 2020) 99
Fig. 4.12 Factors affecting coastal aquifers (after Kumar 2006) 99
Fig. 5.1 Map depicting the distribution of reserved forests and mining lease areas from Goa (after Pascual et al. 2013) 119
Fig. 5.2 Extensive tidal flat used for agriculture, on the southern bank of the river Chapora at Tuem (after Dessai 2018) 121
Fig. 5.3 Rice (paddy) fields rendered uncultivable due to the deposition of particulate matter washed down from the mine dumps (courtesy: R. Gauns) 123
Fig. 5.4 Endangered Zuari river ecosystem with several barge maintenance workshops seen in the background 125
Fig. 5.5 Wells affected by saltwater intrusion at Baga, Bardez *taluka* ... 137
Fig. 5.6 A bailey bridge collapse due to excessive overloading in the vicinity of Codli, North Goa 145
Fig. 5.7 Flow chart depicting the mining activity and its impact on the various components of the environment (modified after Pascual et al. 2013) 151
Fig. 5.8 Extensive, abandoned mine pit filled with water at Costi, South Goa 153
Fig. 5.9 Beach degradation due to rising sea level, Sinquerim, North Goa 167
Fig. 7.1 A seasonal lagoon at Ambaul, popularly known as Cola beach, in South Goa 198
Fig. 7.2 Agricultural land on tidal flat (*khazan*) festooned by mangroves, Zuari estuary at Rachol, South Goa 201
Fig. 7.3 Permanent concrete structures right at the waterfront, north Goa 202
Fig. 7.4 A view of the overcrowded beach at Baga on New Year eve (Courtesy: R. Naik) 202
Fig. 7.5 Luxuriant mangroves on the western banks of the Sal estuary at Cavelossim, South Goa 203

Fig. 7.6 The aerial view of lush mangroves (background) at the Kadamba Transport Corporation bus terminus, Panaji encroached upon by urbanization activity (Courtesy: C. Barreto) 204
Fig. 7.7 Large infrastructure project at the waterfront in North Goa 204

Chapter 1
Introduction

Abstract Goa, a miniscule state with rich natural resources and a bounty of natural beauty, is an admired tourist destination in the country popularly known as the 'Pearl of the Orient'. This erstwhile Portuguese colony boasts of serene beaches, forts, churches and temples, some of which date back to the pre-colonial era.

The economy of the State post-World War II, was primarily dependent on mineral wealth supported in later years by a vibrant tourism industry. However, subsequent to the mining ban, tourism almost by itself shouldered the State economy. Although tourism was intended to be inclusive by integrating all stakeholders, particularly the disadvantaged and economically weaker sections of the society, the fruits of economic development have not truly permeated the local stakeholders. Moreover, massive growth in tourism business has impacted the ecology, economy, resources, environment and the social fabric of Goa. As a remedial measure, the State has adopted sustainability as the basic premise of development.

It is hoped that efficient management would conserve resources, secure growth, kindle technological innovations, leading to employment generation, thereby benefitting the consumer by way of more sustainable products. It is expected that a comprehensive approach of sustainable tourism would be able to achieve sustainable development and mitigate the adverse impact on the State in proceeding towards climate neutrality.

Keywords Goa tourism · Western Ghats Escarpment · Aggradational landform · Delta · Extensional faulting · Sustainable development

1.1 Preamble

Goa is the smallest state of the Indian Union. Geographically a part of Konkan, it is a well-defined region along the western seaboard of Indian peninsula. Over the ages, it has been known variously as Aparanta, Mahaspatam, Sunaparant, Gopakapattanam, Gomant and so on. Historically, being a major trade centre on the west coast, it has been a prized territorial possession. As such, Goa has been ruled over centuries

A. G. Dessai, *Environment, Resources and Sustainable Tourism*,
Advances in Geographical and Environmental Sciences,
https://doi.org/10.1007/978-981-99-1843-0_1

by several dynasties such as the Chalukyas of Badami, Rashtrakutas, Chalukyas of Kalyani and their feudatories, the Kadambas of Goa, the Bahamanis and thereafter, by Adilshahi, at the time of its conquest by the Portuguese in 1510, under whom it remained until its liberation in 1961.

Interestingly, although it is a miniscule State, demographically it is fourth among Indian states with a population of 14.57 million (2011 census), with a growth rate of 8.17% (1991–2011). It has acquired worldwide acclaim primarily due to geopolitical reasons. For one, it is endowed with rich natural resources including water, minerals, soils, forests among others and a bounty of natural beauty. Secondly, it boasts of a high literacy rate (88%), higher life expectancy as compared to the neighbouring states, a good physical infrastructure and resultantly growing demographic density of 394 persons per sq km, a degree of socio-economic advancement with a high per capita gross domestic product (GDP), which is 2.5 times that of the country as a whole, and per capita income more than three times the country's average. The economic survey of the State for 2018–2019 indicated a growth rate of 6.23%; however, these figures need to be taken with caution since a large chunk of remittances from other parts of the country and abroad significantly contribute to the State economy. The State displays a blend of rich composite culture, encapsulating diverse subcultures that have had a harmonious coexistence for centuries. This phenomenon is truly representative of unity in diversity. Hence, it comes as no surprise that the people of the State respect inclusive values such as plurality, tolerance, secularism and religious freedom and generally hold a liberal outlook to life.

The State with an area of 3702 sq km is nestled in the verdant lap of the *Sahyadri* (Western Ghats) mountain range, which covers an area of about 600 sq km, and lies between latitudes 14°53′54″ N and 15°40′00″ N, and longitudes 73°40′33″ E and 74°20′13″ E, in southwest India. It is bounded by the state of Maharashtra in the north and by Karnataka in the south and east. The Arabian Sea, renowned for its scenic beaches, forms its western flank, while the *Sahyadri* mountains provide the eastern backdrop. Panaji is the State capital. Margao, the cultural capital of Goa, is a historical town, also known as Mathagram in pre-Portuguese times, and is the modern commercial hub of the State. Vasco da Gama, the most populous port-town, displays one of the distinctly varied and minimally native social structures, which are nonetheless admirably cohesive. The State remained under Portuguese rule for nearly 450 years until 1961, when it was liberated after a military operation and became a part of the Indian Union. It remained as a union territory for nearly 15 years since and attained statehood on 30 May, 1987.

The State is well connected to other towns in neighbouring states and other parts of the country by national highways and railways which run right across the State both N-S and E-W (Fig. 1.1). It has a natural harbour at Mormugão and two international airports, one at Dabolim and the other at Mopa. Thereby, it is accessible to most other state capitals in the country as well as to several countries of the world.

State Resources and Environment: The economic well-being of a region is primarily governed by three types of resources, the natural resources, the human resources and the interactions among people that could be referred to as the social

Fig. 1.1 Location map of Goa

resources. Among all these, the natural resources constitute the very foundation on which the edifice of economic development rests.

Although the economy of the State was initially agrarian, post-World War II, it has come to be primarily dependent on the mineral wealth, albeit supported in later years (post-1983) by a pulsating tourism industry. As such, these two industries have shaped and impacted the socio-economic life as well as the environmental profile of the State. However, since 2012, subsequent to the mining ban, tourism industry has almost by itself shouldered the State economy. Initially, it was promoted as a non-polluting industry, but in due course of time the shortcomings of tourism have been noticed. Although tourism was intended to be inclusive by integrating all stakeholders, particularly the disadvantaged and economically weaker sections of the local community, the fruits of economic development have benefitted more the international tourism business operators and large corporates that promote hotel and restaurant business. The total revenue generated from tourism activities was over ₹406 lakhs (Economic Survey, 2018–2019). The tourism activity has been responsible for the promotion of infrastructure development, primarily construction, leading to soaring land prices and illegal land deals. Both these activities, more than others, are resource intensive, and hence, they consume natural resources as well as generate large quantum of waste to burden the environment. It is for this reason, to achieve sustainability efficient management of resources is of critical importance. Thus, it is evident that massive growth in tourism business has impacted the ecology, economy, resources, environment and the social fabric of Goa. A small and fairly well-managed State is acutely feeling the prangs of damage due to the overutilization of all resources. Concurrently, it is also an ideal region to study the impact of tourism on environment and how the resource base could be salvaged from further deterioration and degradation.

The growing demography and unregulated tourism growth have placed unsustainable demands on the State's natural resources. As a remedial measure, the State has adopted sustainability as the basic premise of development. It foresees sustainable development through sustainable tourism which forms the major objective of tourism development in the State. The various plans and programmes of the State are designed and implemented towards sustainable tourism development. However, an efficient management of resources would be required to secure growth and kindle technological innovations, leading to employment generation thereby benefitting the consumer by way of more sustainable products. Thus, it is expected that a comprehensive approach of sustainable tourism would be able to achieve sustainable development and mitigate its adverse effects. From a larger perspective, improvement in resource efficiency through better management would enable the State to attain climate neutrality by maintaining global warming below 1.5–2.0 °C as visualized by the Paris agreement.

The intent of this narrative is to produce a reference text on the impact of tourism on the exploitation of earth resources and to suggest possible strategies for maintaining the growth rate, reducing poverty and inequality, simultaneously conserving the mineral and other natural resources and preserving the environment. In later chapters, the brunt of excessive exploitation of natural resources and inefficient management particularly of water, the urgent need for the upgradation of infrastructure for sewage and sanitation, and solid waste disposal are critically examined. Qualitative and quantitative assessment of the waste materials is carried with a view to explore their potential applications in recycling and reuse. This would help at least in a small measure in both resource conservation and extension of the useful life of the natural resource base of the State. Finally, the impact of mass tourism on ecology, energy, environment and the social community is discussed and remedial measures suggested in achieving sustainable development.

1.2 Background of the Mineral Resources

Brief accounts of mineral resources and their geological setup were provided by T. J. Newbold in 1844. The earliest well-documented reference to the regional geology of South India is by Bruce Foote (1876) who described several bands of schistose rocks and assigned them to the 'Dharwar System'. A dozen of such bands are described by Krishnan (1968). The rocks in Goa are the northward continuations of one of the schistose bands mentioned above (ibid.). Later, Maclaren (1906) while describing the rocks from the Castlerock area in the Uttara Kannada (North Karnara) district of Karnataka, adjoining Goa, suggested that the rocks continue westwards in Goa and proposed the name 'Castlerock Band' to these schistose rocks. The Castlerock Band which hosts the iron and manganese ores consists of magnetite- and haematite-quartzites, biotite-quartz schists, phyllites, grey limestones and basic igneous rocks along with the gneisses. In the bipartite classification of Smeeth (1916), the rocks from Goa were assigned to the Upper Chloritic Division of the Dharwar System, as most of the rocks exhibit low-grade greenschist facies of metamorphism. The rocks in the south were included in the Lower Hornblendic Division (ibid.). A summary of their work is available in Pascoe (1950).

But for a thick cover of laterite which caps all lithologies, the region is largely occupied by schistose and granitic rocks (Borges 1948; Pascoe 1950; Dhepe 1954) except for the north-eastern corner of the State, where Deccan Traps (66–65 Ma) lava flows unconformably overlie the Dharwar metasediments. A formal stratigraphic classification of the Goa rocks was presented by Oertel (1958) who divided the rocks from Goa into two groups: (i) lower infraconglomerate group and (ii) upper metalliferous group of schistose rocks. The granitic rocks occurring at Quepem, Chauri (Canacona) and Dudhsagar were considered by him to be intrusive into the schistose rocks.

The presence of iron ores in this tiny State was known from reports of the sixteenth-century Dutch traveller John H. van Linschotten. Several centuries later, the deposits were geologically mapped by L. L. Fermor in 1909 and were described in his classic memoir on the Manganese Ore Deposits of India. Later work on the manganese ores by J. A. Dunn (1942) classified the ores as of lateritoid type (op.cit.). This work was followed by detailed geological mapping by the Geological Survey of India (Gokulam

1972; GSI 1981). Numerous studies on geology, ore petrology, geomorphology and the laterites followed in later years.

1.3 Physiography

The eastern parts of the State consist of high-rise mountain ranges and deeply dissected valleys which form the *Sahyadri* mountain range and constitute the Western Ghats Escarpment with elevation ranging from 200 to 900 m, and a few peaks with elevation of over 1000 m above Mean Sea Level (MSL). The highest peak is known as Sonsogarh with an altitude of 1167 m above MSL and is located in the north-eastern part of the State. The central part of the State consists of isolated hill ranges whereas the western region forms a part of a low-level coastal plain, known as the Konkan.

The Western Ghats constitute a geomorphic feature of immense global importance related to the breakup of Gondwana in early Jurassic and the formation of India into an isolated landmass that ultimately collided with Eurasia during the Tertiary. It forms a continental divide between the west and the east-flowing drainage of the Indian peninsula and has thus played a significant role in carving the landform architecture of the State.

1.3.1 Western Ghats Escarpment

The Western Ghats Escarpment represents the erosional remnants of the western flank of the Deccan Plateau in the north and associated hill ranges of Dharwar metasediments and gneisses further south, that run almost parallel to the west coast. These ranges extend from the Kundaibari pass in Dhule district of Maharashtra (south of Tapti River), in the north and reach up to Cape Comorin (Kanyakumari), in Tamil Nadu, in the south, over a distance of more than 1600 km and rise to over 2000 m above MSL. They form a precipitous scarp, at the foot of which lies the coastal lowland (10–25 km wide) known as the Konkan coast in Maharashtra and Goa, and the Malabar coast in Kerala. The mountains are collectively known as the Western Ghats or Sahyadri Mountain Range.

All along the length, a major escarpment at the top, faces west. The ghats form a major water divide between the west- and east-flowing rivers of the south Indian peninsula. It is an asymmetrical divide, since the topography to the west is deeply dissected with steep slopes and swift-flowing rivers with irregular gradients and V-shaped valleys.

On the contrary, the topography to the east is gently sloping, with meandering rivers which exhibit large deltas at their confluence. The main water divide is not always represented by the escarpment, as the plateau has been indented and dissected by headward erosion from west flowing river systems. The streams flowing west are short and swift due to the steep gradient in the upstream sections, which has resulted in the formation of cascades, cataracts and waterfalls.

The mountainous part of the ghats is about 10 km in length. The northern part of the ghats in Goa consists of Deccan Traps lavas (basalts) especially around Chorla ghat in Sattari *taluka/taluk* (an administrative unit, part of a district) whereas the southern part is dominated by Precambrian granitic gneisses and metasediments particularly around Anmode ghat (Sanguem *taluka*).

These ranges are distinct from the orogenic mountain ranges of plate-collisional origin such as the Himalaya, wherein compressional forces have been responsible to bring about the uplift of rocks largely deposited on the ocean floor. As distinct from this type of uplift, the Western Ghats represent the precipitous edge of a rifted, uplifted and deeply dissected plateau, formed at the continental margin. For a long time, the stupendous scarp was considered to be a fault-scarp, associated with the West Coast fault, that affected the western continental margin of India during Miocene-Pliocene times (Wadia 1919; Pascoe 1950; Krishnan 1968).

Recent geological and sedimentological studies both inland and offshore suggest a post-Oligocene period for the ghat formation. Although faulting along the coast was suggested, little evidence of faulting was available in the field. Auden (1949) showed that the basalts in the vicinity of Mumbai dip westerly and attributed this attitude of the lavas to a monoclinal flexure which he named the 'Panvel Flexure'. In recent years, several evidences of faulting have been documented, both on land and offshore. A detailed study based on multispectral data and field mapping along the west coast between Mangalore and Mumbai (Dessai and Bertrand 1995) has indicated that the 'Panvel Flexure' is constituted of fault-controlled blocks which show varying westerly dips as a result of extensional faulting, block-rotation and tilting.

1.4 Drainage

The drainage is largely westerly, save the River Sal, and comprises seven rivers and their tributaries and two rivulets. All rivers are consequent on the initial slope of the hills or the terrain at the source. Later, they may be guided by faults or foliation or bedding of the rocks over which they flow. In a number of cases, they flow across the foliation or bedding of the rocks which is often seen as anomalous behaviour. Such anomalies are often dictated by the physiographic and tectonic features with which the river had to adjust over a long period of time.

The State rivers from north to south, are Tiracol (Terekhol), Chapora, Mandovi, Zuari, Sal, Talpona and Galgibaga, respectively, with two rivulets Baga in the north and Saleri in the south. Of the major rivers, the first four and the last are perennial rivers, whereas the remaining are ephemeral. All rivers drain into the Arabian Sea. The first four of these originate in the Western Ghats, of which two, namely Mandovi and Zuari, are relatively major as compared to the others. The latter meets the sea at Mormugão where it houses one of the best natural harbours of the country. In the lower reaches, about 20 km from the river mouth, the estuarine channels of both these rivers are interconnected by the 'Cumbarjua Canal' distributary. The river valleys are V-shaped in the eastern hilly region. They broaden towards the west and become almost U-shaped on the coastal plain. Both these rivers are extremely beneficial to the State, since they are navigable in their lower reaches providing a cheap means of transport.

Most of Goa is drained by the Mandovi and Zuari river systems. Five types of drainage patterns can be seen: (i) dendritic, (ii) rectangular, (iii) anastomosing, (iv) trellis and (v) barbed. These patterns are largely controlled by the lithologies drained by the river systems. The metabasalts, granitic gneisses and metagreywackes generally display dendritic drainage pattern. Rectangular drainage is usually controlled by the structure of the area and is well developed at Valpoi, Ponda and Netorlim. Coastal streams exhibit an anastomosing pattern due to the low gradient and the influence of tidal incursions. The streams in the higher reaches of the Mandovi basin in general, and the Dudhsagar valley in particular, display a barbed pattern due to stream-piracy (river capture). This is the result of the high rate of headward erosion, steep gradient and high rainfall, all of which have been aided by strong control from the structural features such as the faults, shear- and fracture- zones. The main water divide thus, is not the westerly facing escarpment, since portions of the plateau have been dissected by the headwaters of the west flowing rivers, e.g. the Mandovi.

1.4.1 Tidal Creeks

The lower reaches of all rivers are affected by the ebb and flow of the ocean tide. The waters in these stretches of rivers have variable salinity and electrical conductivity over the tidal cycle. Such stretches constitute estuaries, which form a transition zone between the river environments and ocean environments. They are subject to the influx of saline waters and riverine influence such as the flow of freshwater and sediments. The inflow of fresh and saline waters provides high level of nutrients, in both the water column and the sediments, making estuaries the most productive natural habitats in the world. All rivers from the State have tidal creeks in their lower reaches as they have drowned valleys, where they join the sea. There are practically no deltas. Deltaic conditions are, however, evident from the 'Cumbarjua Canal' distributary, which is an indication that, at some stage, during Holocene the two major rivers may have perhaps experienced deltaic conditions. The possibility of the rivers heading towards the deltaic stage is remote, as the coast is largely of submergence.

The lower reaches of rivers being tidal creeks, all rivers are navigable in their lower sections, at least for a few kilometre upstream. This has been a boon to the mineral resources of the State, especially the iron, manganese and bauxite ores, since waterways provide the cheapest mode of transport.

All rivers from the State drain from east to west as dictated by the general gradient of the terrain or are controlled by the structural weaknesses of the rocks (Dessai and

Peshwa 1982). These controls are not applicable to the River Sal, whose southerly flow is contrastingly exceptional to the westerly flow of all others.

Sal is a tiny rivulet which has tidal incursions in its lower reaches, and unlike most other rivers, it does not originate in the Western Ghats. Its origin can be traced to the Verna '*plateau*' that forms the lowermost planar surface (20–<100 m above MSL) in Goa. The channel of the river is fairly broad in the lower reaches and is marked by mid-stream bars, 0.5–2.0 m in height and a spit at the river mouth, at Betul.

This suggests that the development of Sal is unlike most other rivers from the State and its temporal existence is very limited. The absence of tidal flats *sensu stricto* and extensive landward extent of wind-borne sediments (~2.0 km) coupled with investigations of sediments from the estuary and river mouth leads to the proposition that it is a desiccating palaeo-lagoon (e.g. Kale 1983) or a tidal creek formed as a result of a spit formation at Betul.

1.4.2 Linearity of the West Coast

The western seaboard of the State extends (N-S) over a length of 104 km. It is broadly rectilinear from Querim (Keri) in Pernem *taluka* in the north to Cananguinim (Quepem *taluka*) in the south except that it is punctuated by three estuarine indentations of the Chapora, Mandovi and Zuari rivers, and several headlands (promontories) prominent among which are those at Aguada, Dona Paula and Mormugão. These headlands have been responsible for the formation of one of the best natural harbours of the country, located at Mormugão. The coast, south of Cananguinim, is punctuated by several headlands, the most prominent being the Cape of Rama (Cabo de Rama).

The rectilinear nature of the west coast of Goa in particular, and of India in general, is in stark contrast to the east coast where a number of indentations in the form of deltas are seen to have shaped the coastal configuration. It is inferred that the linearity of the west coast is related to its dismemberment from the Gondwana in general, and its separation from Madagascar (~80 Ma ago) in particular. During the early Tertiary, the western Indian margin witnessed a major faulting event, expressed as the West Coast Fault, along Precambrian basement trends, represented by the NNW-SSE fracture system. This preceded the separation of India from Seychelles (~65 Ma ago).

The configuration of the present-day western Indian margin is the result of the sinistral *en echelon* (left lateral) fault system related to the Tertiary (Dessai and Peshwa 1982; Dessai and Bertrand 1995). The coastal morphology largely shows evolutionary features (both aggradational and degradational) related to the post-Flandrian transgression as the present sea level was established during the mid-Holocene (e.g. Goudie 1973; Bloom 1978). The Goa coast shows ample evidence of sea-level changes related to post-mid-Holocene.

The continental shelf off Goa is about 80 km wide and covers an area of 10,000 km^2. The shelf break occurs at a water depth of 120 m. Further offshore the continental slope is followed by the continental rise, up to water depth of 3000 m. Beyond 3000 m water depth, the region is included under the abyssal plain. It is customary to divide the shelf zone into three regions: (i) nearshore zone up to a water depth of 10–15 m, (ii) inner shelf zone up to 40–50 m of water depth and (iii) outer shelf up to 50–120 m water depth. The inner shelf varies in width from 35 to 45 km, whereas the outer shelf ranges in width from 45 to 80 km. The inner shelf is largely featureless; however, the outer shelf exhibits two shelf-edge ridges between 85 and 136 m water depth. These have been dated by the radiocarbon (^{14}C) technique. The inner ridge is dated to 7.5 ka and the outer one to 11.0 ka (Nair and Hashimi 1980; Rao and Nair 1992).

1.4.3 Beaches

Beaches are among the most common and best developed aggradational landforms of the State that have been a great tourist attraction and thus significantly contribute to the State economy. More than anything else, it is the beaches on which the tourism industry of the State depends. The entire coast is essentially covered by three rectilinear, continuous stretches of picturesque beaches aligned in the NNW-SSE direction. Two, in North Goa extend from Keri-Arambol (Harmal) to Morji over a distance of 9 km and from Baga to Aguada (Sinquerim) over a distance of more than 7 km. The other in the south stretches from Issorchim-Velsão in the north to Cavelossim in the south, over a distance of more than 26 km with some scenic beaches in the intervening stretch at Arossim, Cansaulim, Majorda, Colva, Benaulim, Varca and Orlim.

In addition to the above, small, sandy, crescent-shaped, bayhead beaches, often protected by rocky promontories, occur at Bogmalo, Baina (associated with the Sada promontory), Vaiguinim, Odxel, Cacra, Nauxem (Dona Paula promontory), Dharvel and Ambaul at Cola, Agonda, Palolem and Galgibaga (associated with the promontories to the south of Cape of Rama). The one at Ambaul (2 km north of Agonda) is a small crescent-shaped pocket beach (popularly known as the Cola beach) where a picturesque seasonal lagoon develops at the mouth of the river, across the Shiroti/Shellap rivulet.

1.4.4 Lakes, Tanks and Reservoirs

Lake can be described as a naturally occurring, fairly extensive body of fresh or saline water, usually surrounded by land on all sides. A relatively smaller waterbody, either natural or man-made, is known as a 'pond'. The majority of standing water bodies from Goa are man-made, primarily for irrigation, by blocking entry of tidal waters. They are locally known as '*tollim*' (singular '*tollem*' from the Konkani meaning *lake*) and could be categorized either as '***ponds***', or more appropriately as '***tanks***'.

There are six *lakes* (tanks) in Tiswadi *taluka,* namely ***Karmali*** *Lake* at Carambolim, sadly the most impacted after the construction of the Konkan railway, the ***Toyyar*** *Lake* at Chimbel, the ***Zor*** at Bainguinim, the ***Bondava*** or ***Bondvoll*** *Lake* at Taleigao, the ***Batim*** *Lake* at Goa Velha and the ***Sulabhat*** *Lake* at Agasaim. The ***Mayem*** *Lake*, a fairly popular tourist destination, is located in the Bicholim *taluka.* Seven large ponds exist in Salcete *taluka,* namely the ***Sarzora*** *Tollem* near Chichinim, the ***Ambulor*** Lake at Verna, the ***Rai*** *Tollem* at Curtorim, the ***Kamlaanche*** *Tollem,* ***Divle*** *Tollem,* ***Vadlem*** *Tollem* and ***Adsuli*** *Tollem,* all around Benaulim; the last two are interconnected. There are two, namely the ***Nanda*** *Lake* at Cacoda near Curchorem and the ***Budbudyanchi*** *Tali* (lakelet) at Vichundrem in Sanguem *taluka.* The waters of the last one give out hydrogen sulphide/carbon dioxide in small quantities, due to the decomposition of pyritic metasediments, hence the name. A freshwater ephemeral lagoon known as the ***Paliem*** *Lake,* occurs at Arambol in Pernem *taluka.*

A number of reservoirs have been constructed after liberation (1961) by building dams and weirs across tributaries of several rivers. The Salaulim (storage capacity 227 km^3), the Tillari (jointly with Maharashtra, capacity 1514 km^3) and the Anjunem (capacity 0.04483 km^3) dams are fairly large, whereas the Amthane, Murmunem, Podocem, Birondem, Chandel and Chapoli are smaller in size. All these cater to irrigation and drinking water supply. Two water barrages are located at Opa (treatment capacity 114 MLD) and Assonora (treatment capacity 50 MLD).

1.5 Climate and Vegetation

Goa experiences a warm, humid yet salubrious, tropical climate that is modulated by the *Sahyadri* mountain ranges. They act as key barriers intercepting the rain-charged monsoon winds that sweep the peninsula from SW during late summer. Broadly, the region experiences two seasons, a wet season from June to September during which the State receives heavy rains from SW monsoons. This is followed by a short spell of showers from NE, known as the retreating monsoon, approximately during September and November (regarded as post-monsoon rainfall). The average annual rainfall varies from 2800 to 3500 mm. However, the higher altitude eastern region, the Western Ghats, receives much higher rainfall exceeding 4500 mm, at places such as Dudhsagar. Average temperature broadly ranges from 22 to 34 °C. The diurnal temperature variation in the wet season ranges from 4 to 8 °C and in dry season from 10 to 20 °C. The dry season extends from November to May. Up to January, the weather is cooler, less humid and pleasant. The humidity through much of the year is more than 60% and ranges from 80 to 90% in the wet season. From February to May, the weather is warm and moist; the humidity in the air reaches as much as 95%.

Most of the eastern region is covered by some of the best non-equatorial, evergreen, forests which are recognized by UNESCO-WHC to constitute a part of one of the world's eight 'hottest hotspots' of biodiversity and endemism. Approximately 60% of the total area is under forest cover while about 36% is categorized as agricultural land. Nearly 62% of the forested area is protected under wildlife sanctuaries and national parks. The evergreens include trees such as *Alstonia scholaris*, *Polyanthia fragrans*, *Gravis asiaticum*, *Spondias mangifera*, *Syzygium cumini*, *Calamus rotang*, several species of bamboos more common being *Bambusa vulgaris* and a number of species of ferns (pteridophytes) among others.

The relatively low altitude western region of the State consists of deciduous forests and agricultural lands, whereas the coastal plain primarily comprises agricultural lands and urban settlements. Among the deciduous trees *Terminalia peniculata*, *Terminalia tomantosa*, *Xylia dolabriformis*, *Lagerstoenia indica*, *Ficus religiosa*, *Ficus benghalensis* and *Bobax malabaricum* along with small trees and shrubs predominate. The important agricultural crops include *Oryza sativa* (paddy, rice) which is the principal agricultural crop; *Cocos nucifera* (coconut palm), *Areca catechu* (areca nut palm) and *Anacardium occidentale* (cashew tree) are other major cash crops. Among other crops, *Saccharum officinarum* (sugarcane), *Eleusine coracana* (ragi, nachni) and several varieties of Phaseolus vulgaris (legumes, pulses) and fruits such as *Mangifera indica* (mango tree) *and Artocarpus heterophyllus* (jack fruit tree) are also cultivated (Dhepe 1954).

References

Auden JB (1949) Dykes in western India. Trans Natl Inst Sci India 3:123–157

Bloom AL (1978) A systematic analysis of late Cenozoic landforms. Prentice Hall, New Jersey, p 510

Borges A (1948) Reconhecimento geológico do districto de Goa. An Junta Misc Geog Investi Colomb Est Geol Paleon Lisboa 3:1–57

Dessai AG, Bertrand H (1995) The "Panvel Flexure" along the Western Indian continental margin: an extensional fault structure related to Deccan magmatism. Tectonophysics 241:165–178

Dessai AG, Peshwa VV (1982) Manganese mineralization in the tropical forest area of Goa, India. In: Laming DJC, Gibbs AK (eds) Hidden wealth: mineral exploration techniques in tropical forest areas. Proceedings of Seminar of Association of Geoscientists for International Development, Caracas, Venezuela, pp 170–175

Dhepe PD (1954) A note on the geology, iron and manganese ore deposits of Goa. Poona Univ J Sci Tech 4:28–29

Dunn JA (1942) Manganese ore. Rec Geol Surv Ind 76:1–53

Fermor LL (1909) Manganese ore deposits of India. Mem Geol Surv India 37:1–371

Foote RB (1876) Geological features of the South Mahratta Country and adjacent districts. Mem Geol Surv India 12:70–138

Geological Survey of India (1981) Geological map of Goa. Government of India

Gokulam AR (1972) Iron ore deposits of Goa. Geol Surv India Bull 37:1–186

Goudie AS (1973) Duricrusts in tropical and subtropical landscapes. Clarendon Press, Oxford, p 174

Kale VS (1983) Neogene and Quaternary geomorphology of the Zuari and Mandovi river basins, Goa. Ph. D. Thesis, University of Pune (unpublished)

Krishnan MS (1968) Geology of India and Burma. Higginbotham Pvt. Ltd., Madras, p 536

Maclaren JM (1906) On some auriferous tracks in southern India. Geol Surv India Rec 34:45–136

Nair RR, Hashimi NH (1980) Holocene climatic influence from the sediments of the western Indian continental shelf. Proc India Acad Sci 89:299–315

Newbold TJ (1844) Notes, chiefly geological across the peninsula from Masulipatnam to Goa. J Asiat Soc Bengal 13:984–1004
Oertel G (1958) A geologia do districto de Goa. Communicação do Servico Geológico de Portugal, Lisboa, pp 1–40
Pascoe EH (1950) A manual of geology of India and Burma, vol 1. Government of India, New Delhi, pp 1–483
Rao VP, Nair RR (1992) A revaluation of palaeoclimatic conditions during the Pleistocene and Holocene from the western continental shelf of India—evidence from petrology of limestones. In: Dessai BN (ed) Oceanography of the Indian Ocean, pp 423–438
Smeeth WF (1916) Outline of the geological history of Mysore, vol 6. Bull. Department of Mines and Geology, Mysore State, pp 1–21
Wadia DN (1919) Geology of India. Macmillan Publishers, London, p 360

Chapter 2
Mineral Resources

Abstract The mineral resources of the State are hosted by the metasedimentary rocks which rest over the Anmode Ghat Trondhjemite Gneiss (3.4–3.3 Ga) and the intrusive Canacona Granite (2.9–2.3 Ga) together forming the basement for the supracrustal rocks. The volcano-sedimentary assemblage belongs to the 'Castlerock Band' which comprises two greenstone sequences, a lower metavolcanics-dominated sequence comprising the Barcem Group and a clastic-dominated sequence assigned to the Ponda Group.

The banded iron formations (BIF) which have supported the State economy for over seven decades occur as bands, reefs and lenses of banded-haematite-quartzite (BHQ) and banded-magnetite-quartzite (BMQ) interstratified within the phyllites of the Bicholim Formation. The associated manganese ores, however, occur within both the quartz-sericite-chlorite schists of the Barcem and the pink ferruginous phyllites of the Ponda Group.

The bauxite deposits are largely confined to the Mopa-Verna tableland (palaeo-plain) which consists of a thick carapace of laterite capping various other lithologies. The bauxite mineralization occurs within the lower horizons of the laterite profile.

The most widespread and better developed soil profiles occur on the coastal plain (< 10 m above MSL) where they have developed over coastal alluvium and granitic gneisses. In southern parts of the State, they are found to occur over laterites that cap the schistose metasediments.

Sand extraction is largely 'in-stream' extraction confined mainly to the estuaries and in a few cases upstream sections of the major rivers. As for the marine shell deposits, the best and the largest occurrence of shells is located between Chopdem and Tuem in the bed of River Chapora. Two horizons of shells (0.5–0.7 and 0.3 m thick) occur within a shell-siltstone. It is exposed on the tidal/mud flat and extends laterally from 300 to 600 m; however, the continuity needs to be ascertained by pitting. These resources provide a means of livelihood to the local community.

Keywords Mineral resources · Iron ores · Ferricrete · Bauxite · Sand and gravel · Marine shells

A. G. Dessai, *Environment, Resources and Sustainable Tourism*,
Advances in Geographical and Environmental Sciences,
https://doi.org/10.1007/978-981-99-1843-0_2

2.1 Introduction

Goa forms the north-western part of the southern Indian shield which is an ensemble of several cratons (Naqvi and Rogers 1987) broadly to the south of the satellite imagery lineament known as the Narmada-Son lineament. In modern literature, this lineament is known as the Central Indian Suture Zone (CITZ) (Radhakrishna and Ramakrishnan 1988), which, in older literature, is referred to as the Satpura Mobile Belt. The latter represents a major tectonic zone, active since the Proterozoic, and is also known as the SONATA (Son-Narmada-Tapti) rift system that marks the eastern arm of the Cambay triple junction (Burke and Dewey 1973). It separates the Bundelkhand and Aravalli cratons in the north from the Dharwar, Bastar and Singhbhum cratons in the south (Fig. 2.1). The latter three cratonic blocks are generally referred to constitute the southern Indian shield. The Dharwar craton in particular, of which Goa forms a part, consists of two cratonic blocks, the Mesoarchaean Western Dhawar Craton (WDC) and the Neoarchean Eastern Dharwar Craton (EDC) (Nutman et al. 1992) which are sutured jointly along the Chitradurga Shear Zone and the Closepet Granite Batholith (2.6 Ga) that forms a stitching pluton with the WDC.

This chapter provides a concise summary of the geological framework of the various types of mineral resources required by the industry in an attempt to examining the sustainability of tourism on the physical, economic and socio-cultural environment of the State.

2.2 Western Dharwar Craton

The Western Dharwar craton, the oldest craton from among the constituents of the southern Indian shield, is delimited in the west by the Arabian Sea which represents the western marginal rift. In the east, it is bounded jointly by the Chitradurga Shear zone and the Closepet Granite Batholith along which it is welded to the Eastern Dharwar craton as noted above. The northern boundary, which is represented by the CITZ, occurs about 400 km to the north of Goa (Dessai et al. 2010) and is largely concealed beneath the Deccan Traps flood basalts. In the south, the WDC is delimited by the high-grade granulite terrain, namely the Southern Granulite Terrain (Ramkrishnan and Vaidynadhan 2008), also referred to by some as the Madurai Block. The boundary between these terrains is defined by a shear belt system known as the Moyar-Bhavani and Palghat-Cauvery shear zones. Fermor (1936) first grouped this Archaean province of Southern India into two regions: (i) the Charnockitic (high-grade) terrain and (ii) the Non-Charnockitic (low-grade) terrain. Goa forms a part of the Non-Charnockitic region.

The WDC stabilized around Palaeo- to Meso-Archaean (3.3–3.1 Ga). It comprises predominantly grey granitic gneisses that cover a major portion of the craton. The granitic gneisses and migmatitic gneisses, which form the Peninsular Gneissic Complex, consist of tonalite-trondhjemite-granodiorites (TTG) gneisses (3.4–3.3 Ga,

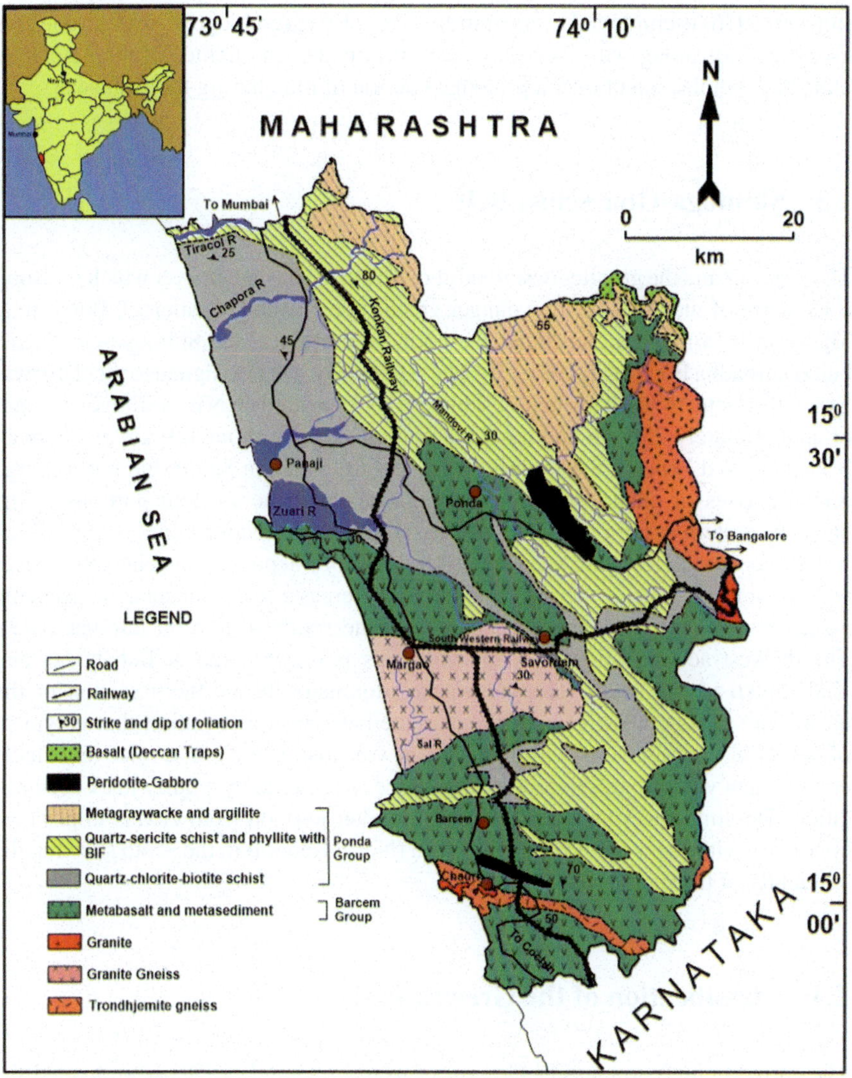

Fig. 2.1 Geological sketch map of Goa (modified after Dessai 2018)

Dhoundial et al. 1987; Devaraju et al. 2007) that crop out in the southern part of the craton and often form the basement for the volcano-sedimentary sequence. The latter comprise the schist belts, the supracrustals (Chadwick et al. 1992), that according to some occur as tectonically transported slices within the gneisses whereas according to others are intrusive into them. In recent years, however, it has been established that the bulk of the schistose metasediments rests over a basement of granitic gneisses and migmatites (Swami Nath et al. 1979; Swami Nath and Ramakrishnan 1981; Ramakrishnan 1994; Ramakrishnan and Vaidyanadhan 2008). The term 'Peninsular Gneiss'

(Smeeth 1916) is employed to encompass a complex assemblage of granitic gneisses and granitoids along with their associated pegmatites and aplites of the basement, indicating emplacement over a prolonged period of time during the Archaean.

2.3 Shimoga-Goa Schist Belt

The gneisses and the granites described above form the basement over which is resting a sequence of metasediments (younger greenstones) of the 'Castlerock Band' (e.g. Maclaran 1906) which constitutes the northern portion of the Shimoga-Goa schist belt (Harinadha Babu et al. 1981; GSI Map 1981; Roy 1983) assigned to the 'Dharwar System' (Foote 1876). This belt of schistose rocks extends NNW-SSE over a length of approximately 250 km and has a maximum width of about 120 km at Dharwar. It is concealed beneath the Deccan Traps and the Arabian Sea in the north. Large domal masses of gneissose rocks and granites mark the western portions of the belt. The eastern margin is faulted whereas the southern section consists of shallow-water deposits. A turbidite sequence (Bouma 1962) deposited in relatively deeper waters comprises the northern part of the belt whereas a subaerial/subaqueous mafic volcanics dominated sequence characterizes the southern portion (Dessai 2018). The thickest accumulation of mafic volcanics is encountered at Bababudan and Kudremukh in the south. The greywackes dominate the northern section of the basin. The rocks consist of an orthoquartzite/basic lava association that is typical of subaerial environments with intermittent volcanic activity in a relatively stable environment with progressive slow subsidence as required by a fairly thick accumulation. The limestone-manganese-iron ore marker horizon from Goa is in physical continuity with rocks from Uttara Kannada (North Kanara) in the south showing the continuity of the rocks of this belt.

2.4 Classification of the Greenstones

The early attempts to classify Goa rocks were by Oertel (1958) who divided the rocks into two groups, the lower infraconglomerate group and the upper metalliferous group of schistose rocks. The schistose metasediments of the Shimoga-Goa schist belt (Pascoe 1950; Krishna 1968) were classified by Gokul et al. (1985) who grouped them into four formations which constituted the Goa Group (ibid.).

In Gokul's (ibid.) classification, all the granitic gneisses (the equivalent of Peninsular Gneisses) and the intrusive granites were considered contemporaneous and younger than the supracrustals (e.g. Harinadha Babu et al. 1981; Gokul et al. 1985). The basement for the supracrustal sequence was not identified. The supracrustal assemblage, however, was tentatively correlated with the Chitradurga Group. The basal metaconglomerate (tilloid ?) was regarded as a para-conglomerate.

This WNW-ESE trending gneiss-granite belt is flanked by the greenstones of the Barcem Formation which rests on a basement of the trondhjemite gneiss (e.g. Dhoundial et al. 1987). The basal conglomerate of this sequence is, however, exposed at Nagargali (Karnataka) associated with quartzites, limestones and other lithologies. The emplacement of the Chandranath Granite Gneiss was followed by the intrusion of basic rocks within the metasediments (supracrustals). The basic intrusives are represented by the Bondla-layered mafic–ultramafic complex and equivalents (e.g. Balakrishnan et al. 1992; Dessai et al. 2009) which were also not adequately represented in Gokul et al.'s (1985) classification.

Thus, there was considerable ambiguity as regards the interrelationship among the granitic gneisses and granites on the one hand, and between the gneisses and metasediments on the other. The supracrustal sequence, in particular, comprises the metavolcanics and clastics each with peculiarities of its own, and despite this, both were included under one stratigraphic group, namely the Goa Group (e.g. Gokul et al. (1985). A revised scheme of classification for the Goa Group was proposed by Dessai (2011). This has been further modified (Dessai 2018) in view of the recent radiometric dates on the Canacona Granite.

2.5 Revised Stratigraphic Framework

In view of the recent work on the gneisses and supracrustal rocks, Dessai (2011) has proposed a revision to the classification of Gokul et al. (1985). In the revised scheme, the Goa Group (ibid.) has been divided into two greenstone sequences (Fig. 2.1), namely:

(i) **Barcem Group** and
(ii) **Ponda Group**

The revised scheme of classification (Dessai 2018) is presented in Table 2.1 and a generalized lithosection for the Goa rocks is provided in Fig. 2.2.

2.6 Palaeoenvironments

The lithological association and the characteristics of the rocks permit to understand the palaeoenvironmental conditions that existed at the time of deposition of the various lithologies. The Barcem Group consists predominantly of mafic volcanic rocks with intercalated clastics. It is characteristic of shallow-water deposition with intermittent subaerial depositional conditions that may have prevailed in the basin. These environments were similar to those that may have existed at the time of deposition of the Bababudan Group. The basal/intraformational polymictic conglomerates and the clastics of the Sanvordem Formation indicate deposition under an active, unstable, shelf environment possibly with intermittent subsidence and upheaval.

Table 2.1 Lithostratigraphic classification of the gneisses and supracrustal rocks from Goa (after Dessai 2018)

Palaeocene-Holocene	Residual Rocks	Ferricretes/laterites
	Basic Intrusives	Dolerites
Palaeocene (66–65 Ma)	Deccan Traps	Basalts
	Metabasic Intrusives	Metadolerites
Proterozoic	Bondla Layered Complex	Dunites-peridotites-gabbros
Ponda Group (2600–2400 Ma)	**Vageri Formation**	Metabasalts Argillites and greywackes
	Bicholim Formation	Banded ferruginous quartzites Manganiferous chert breccias with pink ferruginous phyllites Limestones Pink ferruginous phyllites Quartz chlorite amphibole schists
	Savordem Formation	Metagreywacke Argillite, quartzite, tilloid
Barcem Group (2900–2600 Ma)	**Barcem Formation**	Metagabbros Peridotite, talc-chlorite-schists Quartzite, quartz-sericite-schists Red phyllites Quartz porphyry Metabasalts (massive/schistose)
Archaean (Basement)	Canacona Granite (2900–2300 Ma)	Porphyritic granite
	Chandranath Granite Gneiss (2900–2700)	Tonalite-granodiorite
	Anmode Ghat Trondhjemite Gneiss (3400–3300 Ma)	Tonalite-trondhjemite-granodiorite

The metaconglomerate is characteristic of shallow-water deposition, proximal to the basinal uplifts. On the contrary, the turbidite sequence of the upper part of the formation and the associated volcaniclastics of the Ponda Group represent the distal deep marine facies with intermittent volcanism.

Shallow marine conditions of deposition are indicated by the association of quartzites, stromatolitic limestones and BIF of the Bicholim Formation. The chemogenic precipitation of the BIF is indicative of the terminal phase of sedimentation in relatively deeper parts, under quiet and closed environments. The haematite-magnetite sub-facies variation from north to south is suggestive of platformal deposition. Coexistence of pyrolusite and haematite suggests oxygenated environments whereas fluctuating redox conditions and Eh gradients could be inferred from the association of haematite-magnetite facies transition. The association of black carbonaceous shales is a pointer to the organic activity in ultra-stagnant conditions, intermittent oxygen-deficient environments, at times strongly reducing (indicated

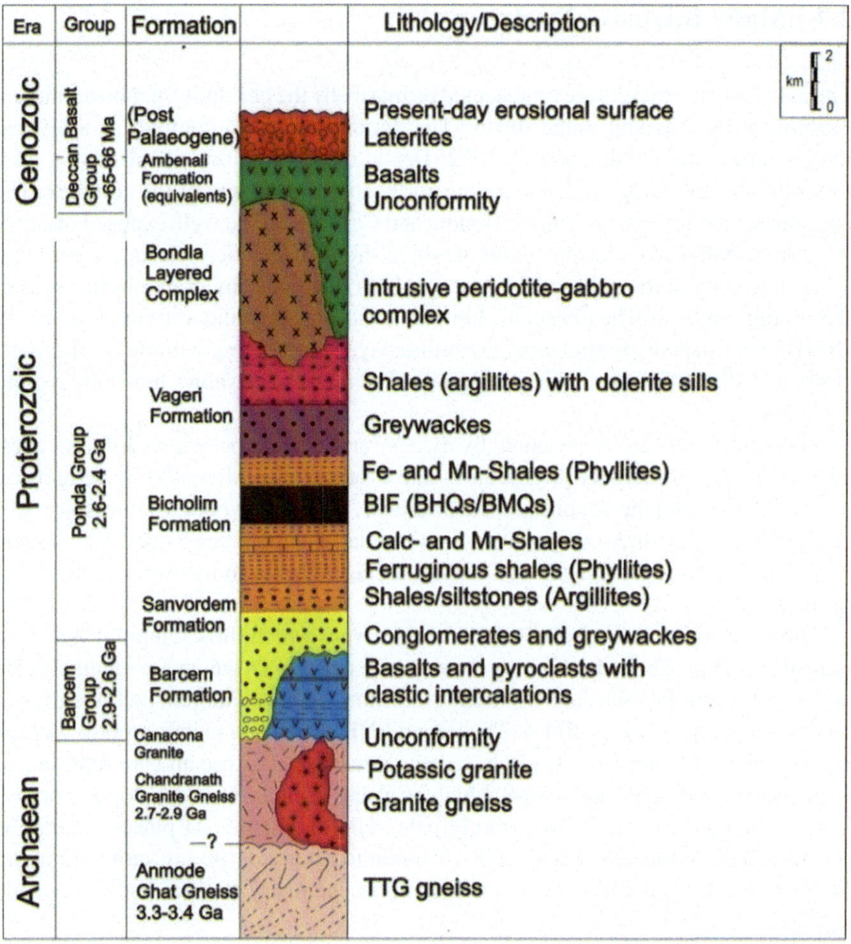

Fig. 2.2 Generalized lithosection for the supracrustals from Goa (after Dessai 2018)

by marcasite), and deeper water conditions that were congenial for the deposition of the sulphide-facies iron formations. The stromatolites within the limestones and cherty silica show possible evidence of prokaryotic microbial activity such as the cyanobacteria which were abundant during the Proterozoic.

Finally, it may be suggested that the Goa basin may have been shallow initially and progressively underwent cyclic deepening as is evident from the flysch facies of greywackes and argillites of the Sanvordem Formation, followed by the shallow-water BHQs of the Bicholim Formation. These in turn are followed by the flysch deposits of the Vageri Formation. The basin remained volcanically active even towards the terminal part of the deposition of the Vageri Formation. The greywackes are suggestive of a provenance dominated by tonalitic gneisses.

2.7 Mafic Intrusive Rocks

The mafic intrusive rocks are represented primarily by the peridotite-gabbro complex, gabbroic intrusives and mafic dykes. The peridotite-gabbro complex is a layered intrusion (e.g. Balakrishnan et al. 1992; Dessai et al. 2009) that consists of at least four plutons, providing evidence for the north–south continuity of the complex. The outcrops occur at Pernem, Valpoi, Bondla and Canacona. The well-exposed ones are those at Bondla and Canacona, both of which show ultramafic rocks at the base that are followed by mafic gabbros, by leucogabbros and rarely by granophyric rocks at the stratigraphic top. The ultramafic horizon at the Bondla mafic–ultramafic complex (BMUC) consists of cumulates of chromitite layers alternating with those of olivine (Dessai et al., 1995), particularly at Bondla. The mafic zone shows uniform layering (op. cit.).

Mafic intrusives are represented by dyke swarms which belong to different ages. The older ones are largely Archaean to late Proterozoic and consist of peridotites, newer dolerites and metagabbros/metadolerites. Among the youngest intrusive rocks can be included the dyke swarms that are terminal to the Deccan Traps. The Deccan Traps unconformably overlie the Dharwar metasediments in the north-eastern part of the State.

The mafic dykes can be subdivided into ***four types*** which show temporal and lithological variation. The oldest among these are the peridotites which are relatively rare and occur at Shirvai, São José de Areal, Netorlim, Aguada, Palolem and Arambol in addition to a few other localities. They trend ENE-WSW, are metamorphosed under low greenschist facies conditions, have been rendered schistose and are represented by serpentine-talc-chlorite schists which at places contain well-developed porphyroblasts of light to dark brown (rarely pale yellow) siderite. At places, siderite is replaced by goethite. Pseudomorphs of limonite after siderite are quite common. Some of the less altered dykes have komatiitic chemistry (Dessai and Deshpande 1979).

The massive (relatively poorly foliated) variety of the dyke rock, which could be referred to as soapstone (steatite), is used to decorate the archivolts (stone-archways) and lunettes of ancient buildings such as the Palácio do Idalcão (Adil Shah's palace—former secretariat of the Goa Legislative Assembly), the Fazenda (Directorate of Accounts), the Indian Customs and Central Excise Museum, all at Panaji. It is also used for masonry work, flooring and even at times for roofing of ancient buildings, for example the Temple at Tamdi Surla. The peridotites are followed by the newer dolerites that are succeeded by metadolerites. Lastly, tholeiitic dykes, namely the dolerites, of Deccan Traps age complete the mafic intrusive sequence.

The newer dolerites are mafic dykes that are subjected to regional metamorphism and hydrothermal alteration. They trend NNW-SSE to almost N-S. They are ophitic to sub-ophitic and are mildly metamorphosed. Clinopyroxene is often uralitized, at places with the formation of actinolite which is altered to chlorite. Plagioclase is labradoritic, invariably saussuritized with the formation of fine-grained albitic feldspar, epidote, zoisite, calcite and sericite. The dykes vary from tholeiitic to calc-alkaline in composition and are quartz-normative with Mg #s varying from 75 to 45. Some of the dykes of this category are completely altered to clay minerals. Such dykes are seen in working iron ore mines, where they cut across the foliation of the banded hematite quartzites. Locally, in the mining belt, they are referred to as 'clay-dykes'. The peridotites and newer dolerites are not described further.

The characteristics of metadolerites and dolerites, which primarily constitute the coastal dyke swarm (Auden 1949), are described below. The metadolerite dykes are quite frequent but next only to the dolerites. They show trends varying between NNW-SSE and NNE-SSW. These are extensively quarried in different parts of Goa for the road metal aggregate which is primarily used by the construction industry.

Dolerites are by far the most common intrusives exposed, especially along the coastal stretch and in the interior, where they intrude the metasediments as well as the granitic gneisses and granites. Although they are sparsely exposed along the length (N-S) of the coast, a swarm of dykes is well exposed at Aguada to the north of Panaji and another at Palolem in Canacona *taluka* in the south. They show trends, varying broadly from WNW-ESE to ENE-WSW, with considerable local variations. In the interior areas, dykes are exposed at São José de Areal and at Tilamol in Salcete and Quepem *talukas,* respectively.

2.8 Iron Ores

Systematic prospecting and exploitation of iron ores in Goa were undertaken for the first time in 1905 by a Belgian mining company named as 'Compagnie des Mines de Fer de Goa' at Bicholim, Sanquelim and Pissurlem. However, the company was forced to close down its operations in India due to the breakout of the World War I (1914). During the period that followed, there were practically no mining operations until the end of World War II (1945). This was followed by a boom period for mining, post-1948, during which both iron and manganese were extensively and almost haphazardly mined on an unprecedented scale with complete disregard to geology, conservation of ores, mine development, mine planning and least of all, the environment, a concept that had not even been mooted in those days. The boom continued during the sixties but gradually waned during the seventies and eighties which was considered the slack period of Goa's mining industry.

The last decade of the twentieth century provided a shot in the arm to the mining industry from the State due to the liberalization of the Indian economy. Over the years, the export market for the iron ores was dominated by Asian countries, namely Japan, South Korea and Taiwan, which preferred ores containing $< 62\%$ Fe. The nineties and the beginning of the twenty-first century provided a boost to the State mining industry due to the export of low-grade ores to China. The Chinese market imported ores that contained even less than 50% Fe, in comparison with the European buyers who preferred ores with more than 64% Fe.

2.8.1 *Ore Deposits: Geological Framework*

The iron ores occur as bands, reefs and lenses of banded-haematite-quartzite and banded-magnetite-quartzite interstratified within ferruginous phyllites of the Bicholim Formation. The ores are controlled by the existence of protores (syngenetic, parent rock) that occur as intercalations within the phyllites at depth. The protore is considered to be a chemical sediment that consists of thin layers of iron oxides, either haematite (Fe_2O_3) or magnetite ($FeO.Fe_2O_3$), alternating with bands of iron-poor, amorphous silica-chert, jasper, tiger eye or shale (Fig. 2.3) and is known to constitute the Banded Iron Formation (BIF). The BIF is sandwiched within the clastics and is underlain by a flysch which has mixed tholeiitic to sub-alkaline volcanics and pyroclastics. It is overlain also by a flysch which has proximal (coarse grained) and distal (fine-grained) clastics intercalated by mafic volcanics. The BIF exhibits ***oxide-carbonate-sulphide iron facies transition***; however, the oxide facies is predominant in terms of distribution, ore quality and quantity.

The iron ores are confined to the Ponda Group which is correlated with the Chitradurga Group of the Dharwar Supergroup (Dessai 2011). They exhibit features typical of shallow marine shelf-facies sediments. The source of iron and silica is believed to be sea water charged with Fe, Mn and Si derived from distal mid-ocean ridge, fore-arc or back-arc hydrothermal activity.

2.8.2 *Structural Disposition*

The BIF extends prominently from Advalpal in the north to Sancordem in the south-east over a length of more than 45 km as a NW–SE trending major fold. The western limb of the fold is best exposed to the north of River Mandovi (Gokulam 1972; Goku, 1985). The closure of the fold is exposed around Sancordem. The eastern limb of the fold is exposed around Valpoi and can be traced further north. The BHQ protores occur at Arvalem waterfall, south of Sanquelim township and at several places around Valpoi. Towards the north, the BHQs continue up to Reddi in Sindhudurg district of Maharashtra, to the north of Goa.

In the south, the BIF extends from Dabal (about 4 km southwest of Dharbandora) in the north and continues southeastwards up to Darguem. Further south, between Rivona and Barazan, it is cross folded along an E-W axis. South of Rivona, the iron ores are closely associated with manganese ores. At Netorlim, and further southeast, the manganese ores predominate almost to the exclusion of iron ores.

The oxide facies is prominent throughout the Northern and Central Goa and continues further north into Maharashtra. The silicate, carbonate and sulphide facies are relatively subordinate and are indicated only from the intercalations within the BIF. The sulphide facies has a very restricted distribution and is represented by pyrite-bearing carbonaceous shale intercalations within BIF at Sirigao (~ 7 km NW

Fig. 2.3 Banded Haematite Quartzite (BHQ), Bicholim Formation, (Ponda Group), Bicholim, Goa, consisting of alternate layers of brown haematite and white cherty silica (after Dessai 2018)

of Bicholim). Intercalations of talc-chlorite-schists occur at Velguem (~ 16 km SE of Bicholim).

Two sub-facies can be identified within the BIF of the western limb of the (NW–SE) fold. The haematite sub-facies occurs in the north and extends from Reddi (Maharashtra) to Bicholim and continues further south approximately up to Darguem (~ 8 km E of Sanguem). Further to the south of Darguem, the magnetite sub-facies predominates and continues roughly up to Barazana-Viliena (~ 17 km SE of

Sanvordem) in keeping with the E-W trending cross folds (to the south of Sanvordem) and the grade of metamorphism.

Along the eastern limb of the regional fold (NW–SE), the magnetite sub-facies attains importance largely to the north of Valpoi. It is well exposed between Dhanoli and Kesari in Sawantwadi *taluka* of the Sindhudurg district of Maharashtra, but is less prominent in North Goa. The magnetic ores of south Goa on an average contain higher percentage of iron (> 25 wt %) whereas those in Maharashtra are relatively impoverished (< 25 wt %, normally between 8 and 23 wt %) being more siliceous than the former.

The BHQ/BMQ band is underlain by manganiferous clays which serve as the marker horizon that indicates termination of the iron mineralization at depth. The manganiferous clay horizon has a total thickness of about 50 m and comprises number of bands, each with a thickness of more than 5–10 m. These are generally underlain by calcareous siltstones or mudstones/impure dolomitic limestones, encountered in drill holes at Sonshi. The BHQs are intruded by the NNW-SSE to N-S trending mafic dykes which are completely altered to clay mineral aggregates.

Environmental Concerns: The production and exports of the iron ore from Goa have progressively increased from about 0.4 Mt in 1951–1952 to over 35.0 Mt during 2008–2009. This quantum of mined-ore has led to the generation of over 120 Mt of overburden during that period (2008–2009). The total quantum of ore exported between 1948 and 2012 is over 450 Mt generating a waste consisting of overburden, low-grade ores and tailings ranging from 800 to 1,000 Mt over a span of 5 decades. This has impacted the environment in various ways as discussed later (Chapters 3, 5 & 7). The other most significant waste from iron and steel manufacture is the slag and gases. Much of the slag is used as an aggregate in cement and related applications. The dust emission is a matter of concern and requires costly particulate collection systems similar to those in coal-based power plants. The global steel making releases about 2.6 Gt of CO_2, which amounts to 6.7% of the world emissions. Most of the emissions are derived from blast furnace reactions and can be reduced only by increasing the energy efficiency of the process.

2.8.3 Reserves

Detailed exploration carried out in 1975 by the Geological Survey of India in North Goa has established that in situ reserves of high-grade iron ores (> 62% Fe) are of the order of 1,061 Mt and are predominantly haematitic. The low-grade ores (> 35% Fe) which are generally of magnetite are placed at 144 Mt. A total of 680 Mt of ore is estimated for south Goa (Harinadha Babu et al. 1981). In addition, fairly large deposits of low-grade (25–40% Fe) magnetite ores occur between Dhave and Sonal. According to the figures available from Indian Bureau of Mines, the total proved reserves are of the order of 367.37 Mt of which 12.48 Mt are of magnetite and the remaining being of haematite (Anon, 2012).

2.8.4 *Production and Exports*

World production of iron ore, amounting to about 3.0 Gt per annum, is dominated by Australia (2.28 Gt), Brazil (~ 0.4 Gt), China (~ 0.37 Gt), India (~ 0.15 Gt) and Russia (0.10 Gt), which are the major producers followed by several others including South Africa, Ukraine, United States, Canada, Iran, Sweden and Kazakhstan among others.

India produced about 0.125 Gt of iron ore in 2014–2015 of which the major contributors were Orissa (50 Mt), followed by Chhattisgarh (25 Mt), Jharkhand (19 Mt) and Karnataka (17 Mt). The production from Goa, subsequent to the lifting of the mining ban, is expected to be about 10 Mt.

Commercial mining in Goa began in 1940s with the export of a few hundred tons of ore. By 1960, exports had touched 7.0 Mt, reaching 12.62 Mt in 1974, and since then, a progressive increase in exports was noticed with every passing year, except for some breaks in between. By and large, from 1990 to 1998 the annual exports fluctuated from around 12.26–17.62 Mt (Anon 2011a, source: Department of Industries and Mines, Govt. of Goa). By the beginning of the millennium, the exports claimed new highs with emerging markets in China. By 2008–2009, total annual exports were over 35 Mt and they touched 53.13 Mt in 2009–2010. The State, however, produced 45.68 Mt of ore in 2009–2010 (Table 2.2). The ore mined in the State is almost exclusively exported. The revenue by way of royalty increased from ₹2920 million in 2008–2009 to ₹5850 million in 2009–2010 (Anon 2011b). Goa contributes about 3.5% of the total Indian export of iron ore which is about 10% of the global annual exports. It must be mentioned that despite a sizable export to the international market, the Goan ores, in general, are not suitable for indigenous consumption due to their inferior quality (i.e. < 62% Fe). The production and export figures of ores presented are for the period prior to the imposition of mining ban in 2012. It may be mentioned that the export figures are far from the carrying capacity of the resource. The State will have to regulate the quantum of export in consonance with the sustainable mining policy to help the State economy and protect and conserve the environment in view of the sustainable development of the State.

2.9 Manganese Ores

Manganese is a strategic mineral resource in most of the industrialized nations of the world. It is one among the most important raw materials of the ferroalloy industry being the essential constituent of carbon steel and high-manganese steel. Manganese steel, also called Mangalloy or Hadfield steel, is a steel alloy containing 0.8 to 1.25% carbon with 11 to 15% manganese. It is non-magnetic steel, very resistant to abrasion, whose surface hardness can be increased by impact without any increase in brittleness. In the steel industry, manganese has two applications. Primarily, it was used as a scavenger to isolate or remove oxygen and sulphur both of which degrade steel.

Table 2.2 Exports of iron ores from Goa (after Pascual et al. 2013)

Period	Ironores from Goa[#] (t)	Ironores from outside Goa (t)	Total exports of Iron Ores from Goa (t)
2002–2003	18,858,230	4,762,732	23,722,880
2003–2004	22,095,993	8,628,647	30,724,640
2004–2005	23,308,033	9,280,138	32,588,171
2005–2006	25,537,924	10,733,726	36,271,650
2006–2007	30,893,953	9,642,721	40,536,674
2007–2008	33,434,429	6,117,626	39,552,055
2008–2009	38,075,223	7,513,548	45,588,771
2009–2010	45,686,900	7,445,102	53,132,002
2010–2011*	46,846,383	389,219	54,424,849
2011–2012*	38,252,554	450,343	43,279,347

#Local production equals to exports; *Tentative figures

This application accounts for 27% of the world consumption. With the advancement in technology, more and more manganese is used as an alloy with iron in the manufacture of various special steel products. About 63% of the world consumption at present is used for this purpose. About 10 kg of manganese ore is required to produce 1.0 t of steel.

Manganese ore of high purity finds application in the manufacture of dry-cell batteries, in which ground manganese oxide acts as an oxidizing agent or depolarizer. Battery grade manganese oxide requires ~ 80% MnO_2 with gamma-MnO_2 (Nsutite) structure, less than 0.05% metals that are electronegative to zinc such as copper, nickel, cobalt, arsenic, lead and antimony, less than 6% iron and no nitrates. Battery grade ore from Gabon has 83–84% Mn, that from Ghana 78%, Greece 75% and India 70% (minimum). Manganese has potential as a component of lithium-ion batteries. It is the main component of chemical compounds such as manganese permanganate that is used to purify water and manganese-organic compounds that are used as fungicides. It is used as an octane-booster in gasoline in addition to applications in paints, pigments, fertilizers, insecticides, bleaching powder and glass among others.

2.9.1 Manganese Ore Resources

Manganese ore deposits of commercial importance occur in the southern part of Goa and have been exploited for over six decades. The first systematic study of the ores was undertaken in 1976 (Dessai 1976). The commercially workable deposits of manganese are located at Rivona, Naveli, Cavarem, Netorlim, Verlem and Salgini in South Goa district. As noted above, manganese deposits are hosted by the *quartz-chlorite schists, quartz-chlorite-sericite schists and phyllites* associated with the *greenstones of the* ***Barcem Group*** (Dessai 2011). In the north, the occurrences of manganese are associated with *quartz-chlorite-sericite-schists and calcareous-ferruginous- and manganiferous-phyllites of the* ***Ponda Group***, although commercially workable deposits are exhausted.

In most sections, the host rocks show extensive lateritization and manganese mineralization is confined to the lateritized portions. The ore occurs as boulders and concretions or as pockets and lenses of varying dimensions within the laterite duricrust. The laterites have in places, retained the foliation of the metasediments despite their alteration to lithomarge. Lenticular bands of manganese, primarily as manganese wad, are exposed at Bhoma to the east of the Sateri temple along the Panaji-Ponda section of national highway NH 748. These are, however, not of commercial importance.

The majority of manganese mine pits show a good correspondence with the major lineament pattern of the area. A photogeological map (Dessai, 1985) showed that most photo-lineaments were due to strike ridges, stream course, valleys and fracture zones. Analyses of aerial photo-lineaments show a group of maxima in the azimuth between N30-60W and N50-60E. The N30-60W trending maxima correspond to the strike of the rocks and the axial plane cleavage of the NW–SE trending folds of the Dharwar orogeny (Deshpande and Dessai 1977; Dessai and Deshpande 1979). The other maxima in the N50-60E direction represent the extension fractures and the axial plane cleavage of the cross folds. These lineaments are more frequent and longer as compared to the others (Dessai and Peshwa, 1982).

***Environmental Concerns*:** Anthropogenic manganese contributions to the atmosphere are small, only about 10–20% of natural contributions, and those to waters are even smaller. Problems with manganese exposure have been observed in some mines, but these are more common in ferromanganese production facilities where they are caused by inhalation of manganese-rich dust and fumes. About 80% of manganese emissions come from iron and steel production facilities. The remaining 20% are from combustion of coal in power plants and coke ovens.

2.9.2 Reserves

Manganese ores of superior quality are localized to Verlem and Salgini, which contain up to 55% Mn. The total recoverable reserves in Goa are estimated at 23.56 Mt (Harinadha Babu et al. 1981). Another estimate by the Geological Survey of India places the reserves at 11 Mt of which 30% are high grade (~ 50% Mn) and 70% are low grade (7–30% Mn). The Indian Bureau of Mines estimates the total reserves at 19 Mt, of which proved are 0.42 Mt (Anon 2012). The production of manganese ore in the State prior to the imposition of mining-ban (2012) was as follows: 2009–2010: 0.77 Mt, 2010–2011: 0.44 Mt, 2011–2012: 1.55 Mt and 2012–2013: 0.05 Mt of 25–35% Mn content (Anon 2012).

2.10 Laterites and Bauxites

The laterites and associated bauxites occur primarily forming tablelands which are closely linked to the palaeoplains. The morphology of the terrain indirectly controls the development of the laterites. From the escarpment in the east to the Arabian Sea in the west, two major planar surfaces can be identified. These are the Anmode planar surface that occurs at the crest of the Western Ghats at an altitude of about 600– > 700 m above MSL and the Mopa-Verna planar surface, about > 10– < 100 m (inclusive of the scattered remnants of 300 m surface), that extends from the foot of the escarpment up to the coast. Three planar surfaces were identified by Dessai (2018). His second surface, namely the Molem planar surface (>150–300 m above MSL), could be considered, a denudational period within the overall development of the Mopa-Verna planar surface.

These surfaces occur as flat, gently inclined palaeoplains formed by subaerial denudation and intense vertical down cutting over a long time, followed by gradual uplift. The repetition of cyclic episodes has resulted in the formation of the step-like pattern of palaeoplains which are blanketed by laterites. They represent the undenuded remnants of elevated palaeoplains.

Although the distribution of laterites in the State is strongly influenced by the morphology of the terrain, it must be noted that the laterite-blankets including the ferricretes, over these palaeoplains, are not indicative of the erosion surface. The laterites may be present on old erosion surfaces and they can also occur completely independent of any erosional surface. Similarly, there is no lithological control on the laterite formation. In general, however, the metasediments exhibit more extensive and deeper lateritization than the granitic gneisses which are invariably stripped off of laterites.

2.10.1 Bauxite Resources

Good quality bauxite occurs on the Mopa-Verna palaeoplain and to a lesser extent on the Anmode palaeoplain. These two altitudinal types can be categorized as the (i) high-level bauxites and (ii) low-level bauxites.

The Mopa-Verna palaeoplain consists of a thick blanket of laterite and forms a tableland (commonly known as the lateritic *plateau*). The bauxite mineralization is largely confined to the lower horizons of the laterite profile. Except for some sporadic occurrences, bauxite deposits associated with the Anmode palaeoplain are not within the jurisdiction of the State. They occur largely in the state of Maharashtra and Karnataka.

(i) **High-Level Bauxites**

The high-level bauxites are largely confined to in situ laterites. Depending upon the lithology on which the laterites have developed, they contain either iron, manganese or bauxite deposits. There is a lithological control on mineralization. Broadly, wherever gneisses are capped by laterites, there is better development of bauxites in comparison with other lithological types. In general, in a vertical section of a bauxite mine, the topmost cover is a hard laterite carapace, the duricrust, about 2–4 m thick which grades downwards into a nodular pisolitic zone 2–3 m thick followed by a massive earthy zone 3–6 m and followed by a granular zone 2–4 m which grades into a clayey zone of variable thickness and gradational into the weathered parent rock. Bauxite is mined from the nodular pisolitic zone and at places, from the massive zone.

(ii) **Low-Level Bauxites**

The low-level bauxites of the Mopa-Verna palaeoplain are localized to (i) Harmal-Keri, Mandrem-Morjim in Pernem *taluka*, (ii) Mapusa-Assagao, Saligao, Porvorim, Verem and Aguada in Bardez *taluka*, (iii) Bambolim, Kadamba *plateau*, Dona Paula, Taligao in Tiswadi *taluka*, (iv) Dabolim, Mormugão, Sada (Vasco da Gama) in Mormugão *taluka*, Cortalim, Verna, Raia in Salcete *taluka*, (v) Canaguinim, Nuem in Quepem *taluka* and (vi) Agonda-Saleri, Galgibag, Loliem, Polem in Canacona *taluka*. All these deposits, however, are not available for exploration and mining as most of the tablelands are occupied either by industrial estates or by other urban/rural settlements. The only tableland where mining for bauxite has been in progress since the late fifties is the Canaguinim *plateau* near Betul in south Goa district. In a typical section of a bauxite quarry at Quitol, Canaguinim, a hard massive laterite 2–4 m thick, grades downwards into a nodular zone of 2–3 m thick, followed downwards by a pisolitic zone 2–4 m which bottoms into a clay zone of variable thickness. The ore is mined from the nodular and the pisolitic horizons. The other potential area for bauxite deposits was the Mopa tableland, in North Goa, where recently an airport is constructed.

Environmental Concerns: Aluminium is extracted from the ore in a two-step process. In the first step known as Bayer process, a solution of caustic soda (NaOH) is used to leach out aluminium from the ore. The dissolved aluminium is then precipitated as $Al_2O_3.3H_2O$, which is calcined to remove water, leaving pure alumina (Al_2O_3). The waste from this process, known as 'red mud', is toxic and hence should be carefully and properly disposed.

In the second step of aluminium production, known as Hall-Heroult process, alumina is dissolved in a 950 °C molten 'pot' of cryolite (a fluoride of Na and Al) and other fluorides of calcium and aluminium. Aluminium is extracted by an electrolytic process during which there are gas emissions consisting largely of CO_2 and perfluorocarbon (PFC) gases, CF_4 and C_2F_6. The fluorocarbons are among the potent greenhouse gases being emitted by this industry. They are about 6,500 to 9,200 times more powerful than CO_2, and they account for as much as 1% of the global anthropogenic greenhouse effects (Abrahamson 1992). However, engineering improvements have reduced the emissions considerably since 1990.

The entire life-cycle of aluminium from the manufacturing stage to the disposal consumes much higher energy than other metals, nearly 7 times more than in steel production. Added to that are the emission concerns and waste disposal. Recycling of aluminium is hence an important aspect of global aluminium industry, since it requires barely 5% of the energy needed to produce the metal from bauxite ore with 95% lower emissions. Nearly 30% of the global aluminium consumption comes from recycled material.

2.10.2 Reserves

The bauxite deposits of North Goa are of better quality than those in the south. The total recoverable reserves are estimated to be 28.09 Mt (Harinadha Babu et al. 1981). As per Indian Bureau of Mines, the proved reserves are of the order of 15.16 Mt (Anon 2012). The Directorate of Mines and Geology, Government of Goa, provides a figure of 70 Mt. The annual production of ore was as follows: 2009–2010: 0.031 Mt, 2010–2011: 0.10 Mt and 2011–2012: 0.084 Mt (Anon 2012).

2.11 Soil Resources of the State

In simple terms, '***soil***' is the upper layered part of the regolith. *It is a natural body of mineral and organic material, that is differentiated into layers, which are related to the present-day surface and which change vertically with depth.* This is in direct contrast to the characteristics of the parent material from which the soil is formed. During weathering, the process of rock decomposition and soil formation indistinguishably merge into one another. A cross-section of soil-outcrop, from surface to the parent rock (bed rock), is commonly referred to as a '*soil profile*'; however, strictly, the term 'profile' refers to an 'outline' and not a 'cross-section'. Nevertheless, the term 'profile' is preferred over 'cross-section', in this text, due to its wider usage.

A soil profile is made up of layers differing in colour, texture and composition. The most important properties significant as far as distinction of layers of soil profile is concerned are the colour, texture, pH, organic matter content, clay mineral type and assemblage, and the amount of sesquioxides such as Fe_2O_3 and Al_2O_3. The ***individual layers,*** which comprise the soil profile, are called '***soil horizons***'. A '*horizon*' is defined primarily with reference to the most obvious physical features, chiefly colour and texture.

The soils from Goa can be broadly categorized as '***Latosolic***' soils or '***Oxisols***' as per the USDA soil taxonomy (Soil Survey Staff 1999). They can be described as '***red residual soils***' or '***ferralitic soils***' (e.g. Duchaufour 1982) formed by leaching of silica and enrichment of aluminium and iron, especially in humid climates. They are characteristic of moist tropical regions. Thickly forested areas, with high temperature and seasonal rainfall, have provided the most favourable conditions for soil development. Free drainage in the upper parts of the soil is an essential feature. Oxisols are typified by deep weathering, extremely thorough leaching, hence contain few weatherable minerals (< 10%), low cation exchange capacity (CEC) and commonly have a marked accumulation of sesquioxides. Kaolinite or halloysite are the typical clay minerals. They often do not have very distinct horizons and are very deep. The differences in properties with depth are so gradual that horizon boundaries are generally arbitrary.

2.11.1 Distribution and Characteristics

The soils from Goa, as mentioned above, have been described here following Duchaufour's (1982) classification and with reference to the physiographic divisions of the State (identified in Chapter 1). The physiographic divisions are as follows: (i) coastal sandy plain and the littoral terraces < 10 m above MSL, (ii) low-level tablelands (*plateaux*) between > 20 and < 100 m above MSL and (iii) high-level Western Ghats tableland/plateau 600– > 700 m above MSL.

The most widespread and better developed soil profiles occur in the coastal plain (< 10 m above MSL) in the watersheds of Chapora, Mandovi and Zuari largely covering parts of Bardez, Tiswadi, Mormugão and Salcete *talukas*. In most of these *talukas*, the soils have developed over coastal alluvium and granitic gneisses. They can be classified as sandy loam to silt-loam and are acidic (pH: 5.5–6.5) in nature, contain moderate organic carbon and are low in potash. In southern parts of the State, for example in Canacona, Sanguem and parts of Quepem *talukas*, the soils are found to occur over laterites that cap the schistose metasediments and are referred to as lateritic soils.

The soils on low-level tableland (between > 10 and 100 m above MSL) are normally very thin (< 15 cm). However, in certain cases, they are formed from alluvium which occurs over the laterites. The profile consists of 'A' horizon-greyish brown in colour (1.4–1.6 m thick) which grades into a yellowish brown 'B' horizon (~ 1.0 m) followed by the 'C' horizon (> 1.0 m) which is often persistent.

In the high altitude (600– > 700 m above MSL) region of the ghats, and in other hilly regions, the soil profiles are thin and poorly developed. Here, the best soil development has occurred over the colluvium on the slopes and in the valleys. High gradient and higher rate of erosion apparently remove the soil formed, allowing little time for a mature profile development.

2.12 Sand and Gravel Resources

Sand and gravel are used as aggregates in the construction industry since long. In recent years with the spurt in construction activity as a consequence of economic development, boost to industrialization and liberalized housing schemes, the demand for these materials has increased manifold. Attendant to that, illegal extraction of sand has also increased significantly.

The beach sand deposits, particularly the inland stabilized dunes, were exploited indiscriminately by the construction industry until the State imposed a ban on sand mining in coastal areas a few years back (e.g. Tiku 1985). Most beaches, in the Central and North Goa, are fairly extensive, except the pocket-beaches from the south which are formed by inland sea-incursions and are protected at either extreme by rocky promontories. The rivers that drain Goa are relatively small. The major ones, e.g. Mandovi and Zuari, flow over a few tens of kilometre before meeting the Arabian Sea. Wherever the hinterland rocks drained by these rivers are granitic gneisses, the beach sand is rich in silica and is hence white. The terrain where the country rocks in the hinterland are metasediments, the beach sands contain accumulations of heavy minerals such as ilmenite, titanomagnetite, magnetite and hematite; rutile, zircon, tourmaline and garnet occur in very small proportions, whereas monazite and thorianite are rare. The concentration of the heavy minerals is low. The rivers drain low-grade metamorphic terrain unlike in the south, in Kerala or in the southeast, along the coasts of Tamil Nadu, Andhra Pradesh and Orissa where the rivers drain high-grade metamorphic rocks such as the khondalite (quartz-garnet-sillimanite-graphite schist or gneiss) and granulites (orthopyroxene + clinopyroxene + garnet-bearing gabbros with a granulitic texture).

Beach sands occur as beach deposits in the inter-tidal zone, recent (ephemeral) dunes on the back shore and stabilized (palaeodune) dunes off the coast, further inland, occur forming ridges parallel to the shoreline (Dessai 2018). The recent (ephemeral) dunes (largely destroyed post-2006) occur within a 10–50 m wide zone and they are devoid of vegetation. The extent of the stabilized dune zone varies from place to place depending upon the topography of the area and the extent of urbanization. In areas where the topography of the coastal zone is flat/plain without undulations and depressions, the stabilized dune zone is quite wide and extends from the recent dune zone up to a minimum of 2 km inland, for instance the coast of Salcete *taluka*. Normally, the first 500 m from the backshore is cultivated with coconut plantations with some rare dwellings. The next 1.5–2.0 km inland is invariably both inhabited and cultivated. It is this zone that was subjected to extensive sand mining in the period prior to the imposition of ban on sand mining.

Present-day sand extraction is largely '***in-stream***' extraction confined mainly to the estuaries and in a few cases upstream sections of the following five rivers namely Tiracol, Chapora, upstream of River Mandovi, e.g. Savoiverem and other places in the vicinity (Fig. 2.4), upstream of Zuari between Borim and Curchorem and the lower stretches of River Sal.

Sand extraction is undertaken on a fairly large scale from the beds of all the above rivers. For example, in the bed of Tiracol within a stretch of less than 2 km in length, more than 100 canoes are engaged in sand extraction. The total extraction is about 3,500–4,000 tons per day. This works out to about 0.9 to 1.0 Mt per year. If the present rate of extraction is permitted to continue, in normal circumstances, the deepening of the river channel could be quite alarming. This could have serious consequences on the rate of erosion and the ecosystem in general. However, in these short coastal rivers under tidal influence, the erosional impact is variable in different stretches of the river basin.

Fig. 2.4 Sand loading activity at a jetty from the Cumbarjua Canal in the vicinity of Amona

2.13 Fossil Marine Shell Resources

The organisms belonging to the phylum Mollusca, the second largest phylum of the invertebrate taxa, include animals such as clams, oysters, snails, mussels and scallops among myriads of others that have a shell or hard outer carapace which protects them from predators and other adversities of nature. A favourable climate with warmer waters, large inter-tidal variations and salinity conditions offers suitable habitats for the growth of variety of molluscs. The present-day shell habitats vary from sandy beaches, rocky shores, mangrove swamps and inter-tidal regions of the estuaries (Sonak 2017). Among the various uses of shells, their use in biostratigraphy and in palaeoclimate reconstruction needs to be specifically mentioned. Additionally, the shells also serve as a useful resource in a large number of industries.

Exploitation of fossil shell-remains of marine organisms, from the river bed, is undertaken on a small scale for manufacture of lime. The term 'lime' is the name of the naturally occurring mineral with chemical composition CaO. The term is also used for both the artificially produced calcium oxide (CaO), also called as 'quick lime', and calcium hydroxide [$Ca(OH)_2$], also referred to as 'slaked lime' (lime putty). Quick lime is usually produced by heating (calcination) calcium carbonate ($CaCO_3$) obtained from limestone, chalk, marble, coral-reef limestone or sea shells at about 1000 °C. Addition of water converts quick lime to slaked lime. Lime produced from sea shells, however, is preferred over that produced from limestone in several industrial applications due to its higher content of CaO. This has led to an increased demand for shells.

In recent years, lime has acquired considerable importance due to a variety of applications in various industries. Lime finds applications in the manufacture of soap (to produce lye or sodium hydroxide), glass, cosmetics, tooth pastes, water treatment, agriculture, sugar and paper manufacture, leather tanning and bleaching, and in mixing cement and concrete.

Concomitant with the increase in demand for lime, the demand for shells also increased particularly for those deposits in the vicinity of urban settlements. It was clear that the resource was becoming increasingly scarce followed by a growing awareness of environmental impact. This was the result of several factors such as (i) pollution of groundwater and surface water bodies due to mining, (ii) degradation of agricultural land in the vicinity of mines due to mine wash, (iii) increase in tourism due to rich biodiversity both on land and underwater and (iv) awareness of global warming as a threat to low-lying areas.

2.13.1 Resource Characteristics

All major rivers from the State contain fossil shell deposits of varying extents. They occur in the inter-tidal zone in the lower reaches of the estuaries, as for example, at Chopdem, Agarwada near Siolim, Moira, Durbhat, Chicalim and several other places. In some cases, localized occurrences of shell beds are noticed in supra-tidal regions also.

The best and the largest occurrence of shell deposit, is located between Chopdem and Tuem in the bed of River Chapora. The deposit is exposed on a tital/mud flat and extends laterally from 300 to 600 m. About 0.5- to 0.7-m-thick clayey-silt is underlain by 0.7- to 1.0-m-thick bed of shells. This is underlain by a layer of clayey-silt (0.20 m) which is followed at depth by another shell bed about 0.30 m thick. The shells occur within a shell-siltstone in which the shells make up about 58 wt %, silt 24 wt % and clay 17 wt %. Silt is principally made up of quartz. The clays consist of minerals belonging to the mixed-layer group of clay minerals.

The maximum area of interest is about 1.8 km in length and about 0.5 km wide. The areal extent is about 11,25,000 m^2. Considering the average thickness of the deposit as 0.8 m, the probable tonnage inclusive of overburden is about 10,80,000 tons. One ton of rock mined, on an average, contains about 580 kg shells. The inferred reserves of shells, therefore, work out to be 6,26,400 tons. If a conservative estimate of the quantity of ore extracted per year is taken as 10,000 tons, then a total of 2,50,000 tons of ore is extracted during the last 25 years. Thus, the area has potential inferred reserves of about 3,75,000 tons valued at ₹168.75 million (lime market value ₹435.7 million). A part of the resource could be categorized as recoverable provided exploitation is undertaken systematically by manual means, observing the mineral conservation measures.

References

Abrahamson D (1992) Aluminium and global warming. Nature 356:484

Anon (2011a) Draft sectoral report, Regional Plan for Goa, Town and Country Planning Department (unpublished)

Anon (2011b) Economic Survey 2010–2011b. Directorate of Planning, Statistics and Evaluation, Government of Goa, Panaji, 146

Anon (2012) Indian minerals yearbook. Indian Bureau of Mines, Government of India, Nagpur, Vol III, Part III, 28–35

Auden JB (1949) Dykes in western India. Trans Nat Inst Sc India 3:123–157

Balakrishnan S, Abbas MH, Vidyadharan KT, Raghunandan KR (1992) Chromite and sulphide mineralisation in the mafic-ultramafic complex of Usgao, Goa. Indian Miner 41:303–322

Bouma AH (1962) Sedimentology of some flysch deposits. Elsevier, New York, p 167

Burke K, Dewey JF (1973) Plume generated triple junctions: key indicators in applying plate tectonics to old rocks. J Geol 81:406–433

Chadwick B, Vasudev VN, Krishna Rao B, Hegde GV (1992) The Dharwar Supergroup: basin development and implications for late Archaean setting in western India, southern India. In: Glover JE, Ho S (eds) The Archaean: terrains, processes and metallogeny. Univ West Aust Publ 22:3–15

Dessai AG (1976) Geology of the manganese ore deposits of Sanguem district of Goa, India. Ph D. thesis, University of Pune (Unpublished)

Dessai AG (1985) An appraisal of the manganese ore deposits of Goa. Proc Indian Nat Sci Acad 51:1021–1032

Dessai AG (2011) The geology of Goa Group: revisited. Jour Geol Soc Ind 78:233–242

Dessai AG (2018) Geology and mineral resources of Goa. New Delhi Publishers, New Delhi, p 323

Dessai AG, Deshpande GG (1979) Mode of occurrence, controls of localization and genesis of manganese ore deposits of Goa. Geoviews 5:21–29

Dessai AG, Peshwa VV (1982) Manganese mineralization in the tropical forest area of Goa, India. In: Laming DJC, Gibbs AK (eds) Hidden Wealth: mineral exploration techniques in tropical forest areas, Proc Semin Assoc Geosci Int Dev, Caracas, Venezuela, 170–175

Dessai AG, Arolkar DB, French D, Viegas A, Viswanath TA (2009) Petrogenesis of the layered mafic-ultramafic complex, Usgao. Goa Jour Geol Soc India 73:697–714

Dessai AG, Peinado M, Gokarn SG, Downes H (2010) Structure of the deep crust beneath Central Indian Tectonic Zone: an integration of geophysical and xenoliths data. Gondwana Res 17:162–170

Dessai AG, French D, Arolkar DB (1995) Geochemistry of stratiform chromites from the Bondla mafic-ultramafic complex, Usgao, Goa. India Jour Ind Assoc Sedimentologists 15:17–29

Deshpande GG, Dessai AG (1977) Occurrence of nsutite (gamma-MnO_2) in the manganese ores of Sanguem district, Goa, India. Curr Sci 46:523–524

Devaraju TC, Huhma HC, Kaukonen RJ, Alapieti TT (2007) Petrology, geochemistry, model Sm-Nd ages and petrogenesis of the granitoids of the northern block of western Dharwar craton. Jour Geol Soc India 70:889–911

Dhoundial DP, Sarkar DK, Trivedi JR, Gopalan K, Potts PJ (1987) Geochronology and geochemistry of Precambrian granitic rocks of Goa, Southwest India. Precamb Res 36:287–302

Duchaufour P (1982) Pedology: Pedogenesis and classification. George Allen and Unwin, London, p 448

Fermor LL (1909) Manganese ore deposits of India. Geol Surv Ind Mem 37:1–371

Fermor LL (1936) An attempt at correlation of ancient schists of Peninsular India. Mem Geol Surv Ind 70:1–218

Foote RB (1876) Geological features of the South Mahratta Country and adjacent districts. Mem Geol Surv Ind 12:70–138

Geological Survey of India (1981) Geological map of Goa. Government of India

Gokul AR (1985) Structure and tectonics of Goa. Proc Semin Earth Resour Goa's Dev, Geol Surv Ind, 14–21

Gokulam AR (1972) Iron Ore deposits of Goa. Geol Surv Ind Bull 37:1–186

Gokul AR, Srinivasan MD, Gopalkrishnan K, Vishwanathan LS (1985) Stratigraphy and structure of Goa. Proc Semin Earth Resour Goa's Dev, Geol Surv Ind, 1–13

Harinadha Babu P, Vidyadharan KT, Jayaprakash AV (1981) Geology and mineral resources of Karnataka and Goa. Misc Publ Geol Surv Ind 30:1–68

Krishnan MS (1968) Geology of India and Burma. Higginbotham, 536

Maclaran JM (1906) On some auriferous tracks in Southern India. Rec Geol Surv Ind 34:45–136

Naqvi SM, Rogers JJW (1987) Precambrian geology of India. Clarendon Press/Oxford University Press, Oxford Monograph on Geology and Geophysics, p 216

Nutman AP, Chadwick B, Ramakrishnan M, Viswanatha MN (1992) SHRIMP U–Pb ages of detrital zircon in Sargur supracrustal rocks in western Karnataka, southern India. J Geol Soc Ind 39:367–374

Oertel G (1958) A geologia do districto de Goa, Communicacao do Servico Geologico de Portugal, Lisboa, 1–40

Pascoe EH (1950) A manual of geology of India and Burma. Govt of India, New Delhi 1:1–483

Pascual XL, Cascallar MP, Planell i Calle M, Rodes EP, Carbonell CS, Vile i Casau A (2013) Iron mining in Goa (India) an interdisciplinary study, Master degree dissertation, Autonomous University of Barcelona, Spain (unpublished)

Radhakrishna BP, Ramakrishnan M (1988) Archaean Proterozoic boundary in India. J Geol Soc Ind 3:263–278

Ramakrishnan M (1994) Stratigraphic evolution of the Dharwar craton. In: A manual of the geology of India, Vol. 1, MGD Centenary Volume, Geol Surv Ind Sp Publ, 77:6–35

Ramakrishnan M, Vaidyanadhan R (2008) Geology of India. Geol Soc Ind 1:556

Roy A (1983) Structure and tectonics of the cratonic areas of southern Karnataka. In: Sinha Roy S (ed) Structure and tectonics of the Precambrian rocks of India. Hindusthan Publ Corp, New Delhi, pp 81–97

Smeeth WF (1916) Outline of the geological history of Mysore. Bull. Department of Mines and Geology, Mysore State, 6:1–21

Soil Survey Staff (1999) Soil taxonomy: a basic system of soil classification for making and interpreting soil surveys. Natural Resources Conservation Service, 2nd edition, U. S. Department of Agriculture Handbook, 436

Sonak SM (2017) Marine shells of Goa: a guide to identification. Springer, Cham, 249. https://doi.org/10.1007/978-3-319-55099-2

Swami Nath J, Ramakrishnan M (1981) Early Precambrian supracrustals of southern Karnataka. Mem Geol Surv Ind 112:1–350

Swami Nath J, Ramkrishnan M, Viswanath M (1979) Dharwar stratigraphic code and Karnataka craton evolution. Rec Geol Surv Ind 107:149–175

Tiku AK (1985) Environmental impact of mining projects with special reference to Goa. Proc Semin Earth Resour Goa's Dev. Geol Surv Ind, 518–520

Chapter 3
Mining Industry

Abstract The major industry that shaped Goa's economy since the mid-twentieth century has been undoubtedly the mining industry which in the later part of the century was supported by the tourism industry.

Goa has witnessed open cast mining for more than seven decades. Nearly one-eighth of the State area is impacted by the mining activity. The quantum of ore exported from the State is over 450 Mt generating a colossal waste of over 800–1000 Mt spread over vast areas. The mining waste has affected the physical environment and ecology in various ways.

The basic concepts of the environmental geosciences, the relationship between the natural resource and environment and the connection between mining and the ecosystem are explained since they form the backbone of development. This is followed by the impact of mining on populated areas such as the potential hazards of mass wasting and slope erosion leading to degradation of large areas of fertile agricultural land. Deforestation is another major cause of concern in this biodiversity hotspot of the Western Ghats with a high level of endemism. Likewise, the pollution of aquatic environments is a cause of concern.

The National Mineral Policy which provides the guidelines to enact appropriate legislation to ensure systematic, planned and scientific mining is described. The salient features of the amended MMRD Act of 1957 to bring it in conformity with the National Mineral Policy of 1993, amended in 2008, and subsequently in 2019 are explained. It makes it obligatory to take care of the natural and social environment so as to avoid pollution, deforestation, degradation and social unrest. The rehabilitation and restoration of mined areas and sustainable mining in the State are discussed.

Keywords Opencast mining · Deforestation · Ore beneficiation · Renewable resources · Resource conservation · Carrying capacity

A. G. Dessai, *Environment, Resources and Sustainable Tourism*,
Advances in Geographical and Environmental Sciences,
https://doi.org/10.1007/978-981-99-1843-0_3

3.1 Introduction

The world is facing a mineral resource crisis, since the resource base is finite. Yet for a fast growing population (albeit slower than in the previous decade), the rate of the mineral consumption is faster than the rate of the population growth. Thus, on the one hand, there is a need to increase mineral consumption and on the other, there are challenges to contain pollution and reduction of carbon footprints to control global warming. This is the dilemma that needs to be resolved. It is in this context that the conservation of environment has to be seen.

The major industry that shaped Goa's economy since the mid-twentieth century has been undoubtedly the mining industry, until the imposition of ban on mining (9 September 2012). However, since the eighties, tourism has also emerged as a predominant industry to reckon with. As such, this chapter focuses on the mining activity in the State and its impact on the environment as well as on the local populace.

The approach adopted has been thus: first, the modern concepts of environmental science are explained; this is followed by fundamentals of mining that are essential to the understanding of contemporary environmental issues. Finally, the impact of open cast mining on the various ecosystems and the environment in general, along with the possible mitigation measures, are discussed.

To reiterate what has been mentioned in the Preface, this narrative concerns largely the peak period of mining activity (prior to 2012) in the State. The chapter depicts the situations, the facts and figures prevailing then.

3.2 Environmental Geoscience

Environmental geoscience is the application of geological information to the study of Earth. This is essential since it affects and is affected by anthropogenic (human) activities that are directed at providing knowledge and understanding of various earth processes. The latter enable to protect human health and safety, preserve the quality of human environment and facilitate prudent use of land and its resources. The study of these aspects is now included under a discipline known as the environmental geoscience.

It is a part of earth science that primarily deals with earth's varied environments and the factors that are responsible to threaten them. An earth scientist thus utilizes the environment as his laboratory and monitors the natural processes such as volcanic eruptions, tsunamis and earthquakes among others, as his experiments. The data recorded during these events are then utilized to understand the causes of these events, formalize hypothesis to explain the phenomena, and at times to predict their recurrence.

Broadly, environment is concerned with the sum total of the circumstances that surround an individual or a community (Keller 1976). For the sake of simplicity, it could be studied in two parts: (i) the physical environment and (ii) the social and cultural environment. Under the first category are included studies pertaining to air, water, gases, landforms among others which influence the growth and development of an individual or a community. The second category is concerned with social aspects such as economics, aesthetics and ethics among others which have a bearing on the behaviour of an individual or a community.

The words *science and technology* can and are often used interchangeably, however, they need to be differentiated from each other. The goal of science is the pursuit of knowledge for its own sake. However, the objective of technology is to create products that solve problems and improve human life. Thus, technology is the application of scientific knowledge for practical purposes. Telephones, computers, satellites, deep-sea submersibles and several other modern gadgets that find applications in our daily life are the result of scientific analysis that has been practically applied to develop new technologies. For example, Thomas Alva Edison's (1848) invention of electric light bulb is based on the principle of electricity.

3.3 Concept of Systems

Years of scientific research have led to the realization that the Earth itself is a complex system which consists of a group of interacting sub-systems. Therefore, the current trend in earth sciences is to study the earth from a new perspective called the ***Systems Concept***. A '***System***' can be explained as a set of interactions separated from the rest of the universe for the purpose of study, measurement and observation. It is a group of subjects and phenomenon that are interrelated and which interact with each other. Careful observations and monitoring of the processes enable us to devise models which could be applied to study the environmental issues. On the basis of these models, it is possible to assess the likely impact of transformations that are seen in nature. It is thus possible to know how a change in one system affects another system or systems and consequently the entire Earth system. As an illustration, contributions of greenhouse gases to the atmosphere have a cascading effect on climate, which has an influence on the hydrological cycle. The latter in turn affects the biodiversity and impacts the lithosphere by bringing about sea-level rise, which affects the coastal regions of our planet.

Modern science visualizes Earth as a single, complex, largely closed system, ignoring the contributions from Sun's energy and the gravitational forces from other celestial bodies. It exchanges very little matter with its external environment in space. The system being complex, it becomes difficult to evaluate and comprehend. In order to simplify the system, it is compartmentalized, for the sake of convenience, into several sub-systems such as the atmosphere, hydrosphere, lithosphere and the biosphere. Alternatively, it can be said ***the earth system consists of four closely coupled components or sub-systems, three of which are inanimate, namely earth (geosphere/lithosphere), water (hydrosphere) and air (atmosphere) and the fourth, the animate, namely life (biosphere).***

Matter and energy are continuously transformed from one state to another through a number of sub-systems. All these sub-systems interact and influence each other, are virtually inseparable and collectively influence the Earth. They constitute an intricate balance, a complex series of interactions in which events in one system exert impact on the conditions in another. The systems approach is a manner of understanding the interconnection between and among systems and how these changes bring about transformations on the planet with time. Since the concepts are devised by humans for their own safety, benefits and comforts, the humans always remain centre stage and thus form the most crucial component of the Earth system. They depend on the Earth for resources, influence the environment and react to the changes in a manner that the latter do not impinge adversely on their existence.

There is increased consciousness among humans of their impact and dependence on the environment and on the interdependence of the different parts of the Earth's systems. All the earth systems are interlinked so that a change in one can affect processes operating in another. The environmental systems are, therefore, to be viewed as interconnected rather than isolated. It is for this reason that a multidisciplinary approach is required to study the Earth processes. In the study of environmental geoscience, ***four issues attain critical importance: (i) population growth, (ii) consumption of earth resources, (iii) hazards due to anthropogenic activity and (iv) pollution and environmental degradation.***

A system consists of two components, namely a ***reservoir*** and a ***flux***. A ***reservoir*** could be visualized as a receptacle in which a particular material resides. The contents of the reservoir are known as the ***stock*** which may consist of either matter or energy. The matter and energy that move from one reservoir to another are known as the ***flux***. It is the rate at which the material and/or energy moves from one reservoir to another and can be measured in quantity per unit time. The term ***sink*** refers to a reservoir and the amount of fluid that it can hold in store is its ***stock***. The rate at which the fluid flows into and out of the system is its ***flux***. For example, consider the movement of water through the atmosphere. Evaporation fluxes add water to the atmospheric reservoir from the surface of the land and ocean. Similarly, precipitation, as rain and snow, delivers water back to the surface of the Earth. These are inputs and outputs to the atmospheric water reservoir.

One of the advantages of the systems approach is, it is possible to learn the behaviour of the system and therefore, the manner of maintaining the balance by measuring the flux of energy and matter moving into or out of a system and from one reservoir to another. For example, an ocean is a carbon dioxide sink. CO_2 migrates from the atmosphere into the water and chemically reacts with water molecules to form carbonic acid (H_2CO_3) which undergoes partial dissociation in the presence of water to yield (H^+) and (HCO_3^-) ions. The increase in (H^+) (hydronium) ion concentration increases the acidity of the oceans known as ***ocean acidification***. As carbon dioxide is quickly processed in the sea, the CO_2-holding capacity of the oceans is ten times higher than that of freshwater. Ocean absorbs a large part of the anthropogenic carbon dioxide which is about 10% and represents significant proportion of the dissolved inorganic carbon in the ocean. It contains about 38,000 Gt (gigatons; 1 gigaton = 10^9 tons) of carbon, which is 16 times that of the terrestrial biosphere. The ocean is therefore, known as the greatest of the carbon reservoirs and essentially determines the atmospheric CO_2 content.

3.3.1 Types of Systems

Systems could be classified into three types: (i) ***Open***, (ii) ***Closed*** and (iii) ***Isolated*** depending upon the freedom by which matter and energy can move across, from one system to another.

An ***Open system*** is one, which permits free movement of both matter and energy in and out of the system. For example, an ocean is an open system, as it allows matter (water, gas, sediments) and energy (from the sun) to enter and leave the ocean. Open systems are the most common type of systems in nature.

A ***Closed system*** permits to exchange energy across its boundaries, but does not exchange matter. No naturally occurring system on the Earth is truly closed. However, systems are presumed to be closed or isolated in order to simplify the models and calculations. Earth as a whole, as mentioned above, is sometimes treated as a ***closed system*** since very little mass enters or leaves over long time spans of millions of years, except for the occasional meteorites and minute quantities of cosmic dust that enter Earth's atmosphere and small quantities of light elements that escape from the outer atmosphere. But for these, Earth's mass has remained relatively constant for over 4.5 Ga.

An ***Isolated system*** does not allow either energy or matter to cross its boundaries. That means, an isolated system does not interact with its surroundings. For example, a controlled laboratory experiment represents an isolated system.

It is also customary to categorize the systems on the basis of the work done, that is, the changes observed in the system with time. Accordingly, the systems could be grouped as ***Dynamic*** and ***Static***. In a ***Dynamic system,*** energy is utilized to bring about a change in the state of the system with time. Most Earth systems are dynamic. Life is a dynamic system which is sustained by the energy from the Sun. However, a ***Static system*** is one in which no change in state of the system occurs, as no work is done. A good example of such a system is the Moon, in which negligible changes if at all, occur with time. The ability of the system to bring about a change is known as the ***energy*** of the system. The mode by which a change is effected in the system is known as the ***process***. The natural processes such as orogeny, tectonism and volcanism among others result in changes to the environmental systems.

3.4 Relationship Between Natural Resource and Environment

Natural resources are the materials and components that occur within the environment. They constitute the backbone of development. No development is possible unless natural resources, whether renewable or non-renewable, are engaged with and put to use for consumption and utilization. The resources being 'natural' by definition, their exploitation has to affect 'nature' in some way or other. In other words, exploitation and utilization of the natural resources affect the immediate surroundings in which the humans thrive. Alternatively, it could be mentioned that the consumption of natural resource affects the environment.

No natural resource can be put to effective use without disturbing the environment and no development can occur unless natural resource is utilized. So the exploitation of natural resource and protection of the environment are two facets of the same issue. Both are essential for the betterment and development of human kind/civilization. The environment and development are complementary issues, and therefore, development planning should take into consideration protection of the environment which includes air (atmosphere), water (hydrosphere) and earth (land-lithosphere) in addition to all other major and minor components that affect the health and well-being of humans. Implicit, therefore, is the fact that ***development always depends on the environment and it should be our endeavour to protect and preserve it at all costs***.

As humankind progressed, it realized that the environment in which it lives is getting adversely affected due to the developmental activities such as urbanization, industrialization and all other activities that are necessary for its very survival. Hence, an increasing need is felt to protect and preserve the environment and its biological diversity, namely the ***ecology***. ***Ecology is the study of the intimate relationship between organisms and their environment, which involves the lithosphere, hydrosphere, atmosphere and the biosphere*** (Tank 1983). The roots of the concept of 'environment' lie in this awareness.

3.5 Global Efforts at Protection of the Environment

First systematic and collective effort by the global community to voice its concerns for the environmental protection and conservation was made at the first Human Environment Conference held at Geneva in 1972. The initiatives taken at Geneva were forerunners for the many initiatives by the world governments including the one at the Rio summit in 1992. At this conference, it was apparent that there has been a global change in international thinking and there seemed to be a concerted effort to protect the environment, which meant to safeguard the future of the generations to come. This concern was expressed in agenda 21 of the conference, a commitment for '***green environment***' that could be made globally safe to live-in and which was adopted by all the nations of the world. Since these early efforts, the concern for the environment has gone a long way. The matter forms a part of the agenda of most of the national and international fora being held on a regular basis, since protection and preservation of the environment are crucial for the survival of human race. Worldwide non-governmental organizations have proliferated as saviours/protectors of the environment.

Education is now focused towards understanding the environment and to find out ways and means to conserve it. The concern for the environment is so acute that most governments have treated the study of environment as a policy in the education curricula. There is demand for certain accountability for disproportionately utilizing the environment for unconscionable profits, as can be seen from the 'polluter pays' principle and the principle of 'common but differentiated responsibility'. Unfortunately, in the developing world, the awareness/propaganda to protect the environment has been lopsided. On the one hand, the need to preserve the environment has been made to appear so severe, that it is being skilfully utilized by some vested interests as a means of livelihood, often with the blessings of the developed world community. On the other, the concern for the environment is dismissed as nay-saying by those who seek to be extractive of nature to the point of no return, to legitimize the over-extraction. Environment has in fact been a convenient platform for the politicians the world over to play power politics/grab power.

3.6 Modern Concepts and Ideas

As more and more attention is being paid to the environmental issues, new ideas and concepts hitherto unheard of are being introduced to explain the regional as well as global changes that affect the environment and ecology. Although most issues that are being discussed are general in nature, and do not pertain directly to any individual per se, they are all indirectly concerned with collective functioning of the communities. In that sense, every individual is indirectly involved and has a role to play.

As development and environment are at cross roads, all developmental activities have some impact on the environment which is directly felt by the local population. Hence, the current trend is to seek development, taking care not to disturb the environment beyond the essential minimum. A new concept of ***eco-friendly sustainable development*** has thus emerged in recent years.

Development is always supposed to be for improvement of quality of life. It is an attempt to improve the socio-economic conditions (by providing adequate resources to the region in question) and the resource base of the region in question. The developing and the underdeveloped societies, particularly of the third world constituted of Africa and Asia, suffer from one great menace, that is poverty. One of the factors responsible is population explosion, which imparts stress on the availability and the utilization of natural resources. The development is fast paced and often without adequate planning which, therefore, adversely affects the natural resources of the region. Thus, sustenance and development always go together, hand in hand, and they cannot be studied as isolated entities.

3.6.1 Sustainable Development

Sustainable development is related to a finite number of users that utilize any resource and finite number of uses that any resource is put to. The ***threshold number of people, beyond which further additions cannot be sustained, is known as the 'carrying capacity'***. If we consider Earth as a resource, it can sustain a fixed quantum of human population with clean air, clean water and adequate nourishment for its survival (e.g. Cohen 1995). Thus, ***the number of people and the quantum of resource necessary for their survival are the qualifying factors that define carrying capacity***.

Resource can be explained as anything that is useful for the needs and requirements of humans, and which is obtained from the environment. Air, water and biomass (plant and animal matter) are resources available from the environment. Other resources such as mineral resources including crude oil, various metal ores such as iron, copper, lead, zinc, silver, gold and groundwater among others, can be obtained solely because of the technologies available for exploration. By and large, developed countries use more resources than essential for basic survival as compared to the underdeveloped and developing countries. For example, United States which makes up barely 4.8% of the world population consumes over 33% of the world energy and mineral resources.

Resources can be classified under three categories depending upon the degree of renewability as ***Renewable***, ***Non-renewable*** and ***Perpetual***.

Renewable resources also called non-conventional resources are those that may get depleted in short time due to rapid consumption, but can be replenished in long term by natural processes. The maximum rate at which a potentially renewable resource can be utilized without the reduction of its potential for renewal, is known as its ***sustainable yield***. If the threshold of sustainable yield is exceeded, then the resource may get exhausted. For example, a groundwater resource can get depleted by exploitation or overexploitation; however, it can be replenished over time, if managed properly. Another example of a renewable resource is agricultural resource. Likewise solar energy, wind energy, bio-fuels (ethanol, biodiesel), hydropower including ocean and tidal energy, geothermal energy among others, are also included under the renewable category. Most of these release nil or far less greenhouse gases than fossil fuels and hence are also referred to as the green energy resources. Recently, 'Green Hydrogen' that is hydrogen produced by electrolysis of water using renewable energy is also emerging as an alternative fuel—the most environmentally friendly fuel. In comparison, 'Grey Hydrogen' which is derived from natural gas or methane, through a process called 'steam reforming' (often referred to as SMR—Steam Methane Reforming), releases carbon dioxide (CO_2) emissions, though these are less than those released in producing 'Black or Brown Hydrogen' which uses bituminous coal or brown (lignite) coal in the hydrogen-making process.

Non-renewable resources are those that are finite and exhaustible, if not managed efficiently, for example minerals, fossil fuels such as coal, petroleum and natural gas, among others, are finite, non-renewable, non-replenishable and exhaustible. They are, therefore, known as '***wasting assets***'. New mineral deposits are forming, but the rate of formation is extremely slow involving millions of years. That is, they cannot be replenished on the scale of human lifetime. For example, formation of oil requires that the biomass (plant and animal matter) is buried for millions of years, is compressed and heated under specific geological conditions. The rate of oil formation in this case is so slow, that the resource cannot be considered as renewable. Some of the non-renewable resources especially those that are extracted from mineral materials, such as the metals: iron, copper, aluminium and lead among others, can be reused or recycled to conserve the supplies. The recycling process involves collection, melting and reprocessing. In contrast, the fossil fuel resources cannot be recycled, since once used they get converted to ash and fumes that go into the atmosphere and heat, that leaves the earth as low temperature radiation.

Perpetual resources are those that are inexhaustible or infinite on human time scale, which involves decades to centuries. For example, solar energy is instrumental in fuelling (insinuating) numerous reactions that ultimately culminate into life on earth. The process is on for the last 4.6 Ga and will continue much beyond the extinction of life on this planet. Other examples are, heat energy from the interior of the Earth and wind energy generated by atmospheric processes (Fig. 3.1), among others.

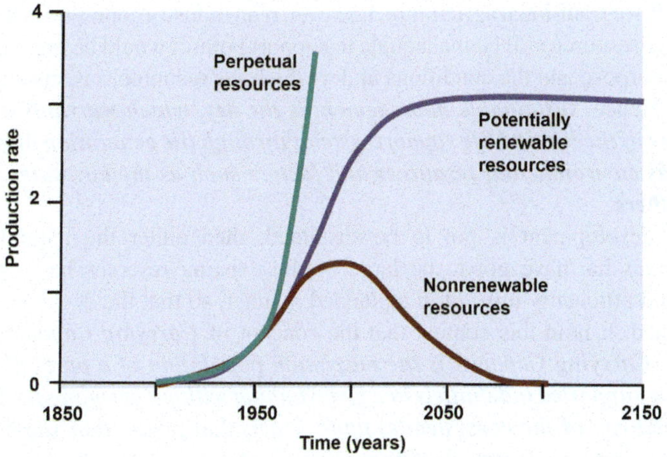

Fig. 3.1 Relationship of the perpetual, renewable and non-renewable resources with time. Perpetual resource can increase continuously with time. The non-renewable resource tapers with time. Renewable resource can continue provided the rate of production is identical to the rate of renewal (modified after Merritts et al. 1998)

3.6.2 Sustainability versus Growth

The rate of development and growth is directly proportional to the rate of exploitation of natural resources. The prime source is the natural mineral wealth. Therefore, faster the rate of exploitation of minerals required for the industry, faster will be the growth/development. However, the mineral resourcebase being finite, the need for judicious exploitation is essential, so that the development is controlled and therefore, systematic. Here lie the seeds of ***sustainable development***.

The World Commission on Environment and Development (WCED 1987; later known as Brundtland Commission), Our Common Future, has defined sustainable development as the one '***that meets the needs of the present generation without compromising the ability of future generations to meet their needs***'. The concept of sustainable development envisages the protection of natural environment and its resources through appropriate management strategies. Although the word 'sustainable' has been adequately explained, the process of achieving sustainable development has not been clear to many, to make the world commission report operational (Munn 1992). A number of questions have been raised. The prime question is, sustainable for whom? Sustainability is relative, therefore, whatever is sustainable for at the level of subsistence is not sustainable at the level of luxury. Therefore, ***sustainability in the context of humans will vary depending upon the socio-economic conditions of the communities involved***. Each individual or a community will have different levels of sustainability. The ***sustainability will also change depending upon the purpose and the conditions***. For example, a groundwater resource in an area may be sustainable for traditional agriculture; however, if an industry comes up in the area, the same water resource will be inadequate to support both. It would be non-sustainable, since the purpose and the conditions under which the resource was sustainable have changed. Hence, ***sustainable management is the key, which warrants a complete knowledge of the life and life support systems through the generation of knowledge on various environmental resources and factors such as the air, water and earth among others***.

If the development is got to be sustained, then either the natural resource base/reserves has/have got to be large or the existing reserves have to be judiciously and efficiently utilized in a planned manner, so that the development could be sustained. It is in this context that the concept of ***Carrying Capacity*** has been proposed. ***Carrying Capacity is the maximum population of a particular species that can be supported indefinitely in a given habitat without permanently impairing the productivity of the ecosystem(s) upon which that population subsists***. As far as human society is concerned, carrying capacity can be explained as the maximum rate of resource consumption and waste discharge that can be sustained indefinitely in a defined planning region, without progressively impairing bio-productivity and ecological integrity.

Vulnerability is the study which determines whether a region or a resource is subjected to potential damage as a result of exploitation. It is the measure of either the ease or difficulty of the damage to a region or resource due to exploitation. Stated another way, it is the ***'measure of insulation' that natural and man-made factors provide a region or resource to be free from degradation***.

Threat, on the other hand, ***is the categorization of the agent responsible to bring about the damage***. For example, a lonely house located at the outskirts of a town or village is vulnerable to burglars. The burglars serve as the threat to the house. Similarly, as far as mining of natural resource in a forested area is concerned, the forest is vulnerable to the damage caused by mining. Mining activity, therefore, serves as the threat to the forest and the environment. Global warming, about which so much is being said the world over, presents a potential threat to civilization. For example, global warming leads to melting of ice which results in sea-level rise, and consequently, the coastal low-lying areas are threatened. The coastal regions are vulnerable to the sea-level rise. Rise in sea level due to the melting of ice is a threat to the coastal population.

Resilience is the capacity of a social-ecological system both to withstand the shock or perturbation either natural or anthropogenic and to rebuild or recuperate itself later. Stated another way, ***it is the capacity of an ecosystem to respond to a disturbance or perturbation by resisting damage and recovering from the damage***, if any. Proper management enables the resource to attain resilience, so as to be able to adapt to the change, and continue to perform and develop.

In view of the vulnerability of the coastal regions to the threat of sea-level rise, the Government of India has initiated a process to formulate strategies to meet the challenges of sea-level rise. One of the initiatives has been to prepare vulnerability zonation maps of the coastal areas which would serve as the basic data required to combat the threat of sea-level rise.

A geological ***Hazard is a naturally occurring phenomenon with the potential for disaster***. For example, huge rock boulders perched at the top of a steep scarp or hill slope constitute a hazard. A destructive avalanche, landslide or rock-fall (when it occurs) is a disaster. Just as there are natural disasters, human-made disasters also occur. For example, oil-spills, air-pollution from the smoke of thermal power stations, factories and boilers. Similarly, a weak dam is a potential hazard for the city/town/settlement downstream. A breach in the dam may lead to a massive flood which is a human-made disaster. The magnitude/potential loss of life and property in the settlement downstream is the potential risk. For example, on 12th July, 1961, heavy rains in the catchment triggered the burst of Panshet (Tanajisagar) dam on the Ambi river (tributary of Mutha), in the very first year of water storage. It resulted in an estimated loss of nearly 1000 lives, in addition to destruction of property worth millions of rupees, due to flooding in Pune city downstream. The breach occurred due to the lack of RCC (reinforced concrete cement) strengthening to the conduit of this earthen dam.

Earthquake and volcanic eruptions are natural disasters that are caused essentially by processes within the Earth's interior. These are often associated with movement of the earth's outer shell known as the lithosphere (100–150 km thick) over the plastic lower layer, the asthenosphere. The lithosphere is segmented into a series of blocks known as the plates that move relative to each other. As the plates move past each other, a part of the plate may get destroyed, whereas another, may undergo growth elsewhere, due to addition of material from within, which is known as accretion. Such large-scale processes of internal origin (endogene processes) which lead to destruction and accretion of plates are known as tectonic processes.

As distinct from the natural disasters discussed above, there are other natural disasters such as floods and landslides which are due to the surfacial processes (exogene processes) that may be related to the more fundamental disorder/disturbance in the interior. Such disasters are often worsened by human interference (activities). For example, deforestation of hill slopes increases run-off of rainwater, thereby enhancing flooding in the streams and rivers downstream. Removal of vegetation also destroys the roots that hold the soil on the slopes. Loosening and removal of soil lead to debris-slip downhill, causing landslides.

Risk refers to the magnitude of potential loss of life (death), injury and the loss of property due to the hazard. The risk of death in India each year by earthquakes is much less than the risk of death by railway accidents or automobile/road accidents or fire. In Northern India, in the Himalaya, the risk of death by earthquake is greater than that in the southern peninsular shield. In fact, much of the Himalayan region of north India is considered to be a high-risk area for earthquakes. A broad belt in central India known as the Narmada Tectonic Zone extending from Kutch in the west to the Himalayan foothills in the east, as well as the western coastal belt of India, has been devastated by major earthquakes. Some of the devastating earthquakes witnessed by peninsular India largely to the south of Narmada in recent times are the Koyna (1967), Latur (1993) and Jabalpur (1997) earthquake among others.

3.7 Environment vis-à-vis Mining: Indian Context

Agriculture and mineral resources constitute the two prime necessities required for prosperity. India has both these in abundance, and coupled with these, it also has an educated, technically trained and skilled human resource base required to scale greater heights in development. In addition, it also has a fairly well-developed infrastructure, political stability, a well-organized banking system and dynamic financial institutions which can provide the desired support system to the growing economy. With the opening up of Indian economy towards the later part of twentieth century, India has been a hub of foreign investment, and consequently, it is one among the leading economies of the world today.

The country is endowed with rich mineral deposits which are spread over wide geographic regions of varied climatic conditions. They range from permafrost in the north to hot and humid in the south, and from desertic in the west to wet and humid in the east. The climatic variation is largely responsible for the variation of vegetation cover, which is expressed by sparse to densely vegetated and forested areas, some of which are ecologically diverse. In most cases, the occurrences of mineral deposits coincide with forested areas. In such cases, there is always a conflict of interests and this at times exerts constraints in both the exploitation of the resource and the protection of the environment.

3.7.1 Measures at Protection of the Environment

Exploitation of mineral resource and preservation of environment both are vital for the existence and sustenance of humans. There can be no industry without the support of mineral raw materials, exploitation of which cannot be undertaken without disturbing the environment. So the situation is similar to the adage 'one cannot have the cake and eat it too'. Balancing between mining and environment has been a challenging task. The National Mineral Policy of the Government of India announced in 1993 is in a sense an outcome of this balancing act and a step forward in bringing about economic and industrial reforms in the country. The formulation of this document is a significant move towards integrating the Indian economy with other global economies.

3.7.2 Mineral Resources

Mineral and fuel resources are natural concentrations of mineral materials that have aggregated at a very slow rate during the long history of the Earth. New deposits are forming at the present time too, but the rate of accumulation is so slow that by the time they become exploitable, the civilizations may perhaps cease to exist. To highlight the slow rate of formation of mineral deposits and the rapid rate at which they are consumed, a corollary with the age of the Earth would not be out of place. Earth is about 4.5 Ga old. Mineral deposits have formed during the last 3.0 Ga. Life began about 1.0 Ga ago. Civilized humans appeared a few thousand years ago. So humans have been using mineral deposits for 0.000 001% of the Earth's history and yet many of the existing ore and fuel reserves are on the verge of being exhausted.

Formation of new deposits requires thousands of years, whereas a large deposit can be exhausted in a few decades. ***Mineral deposits are finite, non-renewable and non-replenishable***. As mentioned earlier, they are, therefore, known as '***Wasting Assets***'. It is for this reason that governments of all countries formulate and lay down certain guidelines for consumption and utilization of mineral resources and it is mandatory that these are followed in all activities related to exploitation of mineral and ore deposits. These guidelines constitute the mineral conservation policy of the country.

3.7.3 Mineral Conservation

In the field of Geology, 'mineral conservation' can be explained as 'exploitation and optimum utilization of mineral deposits by the present generation in a judicious manner so as to avoid wastage'. However, treating minerals as a '***Resource***' involves the human element, since resource is '***anything that is useful for the needs and requirements of humans, and which is obtained from the environment***'. Therefore, mineral resource conservation has a social dimension inclusive of the '***needs and aspirations of the stakeholder***'. Hence, resource utilization should be such that the community depending on the resource should be able to utilize it for as long a time as possible. This, in other words, means optimal use of the resource by present generation without unduly disturbing its potential to meet the requirements of future generations. Thus, the resource should be able to meet the requirements of the future generations too, for the production of goods and services, for which it depends on the resource. Utilization of the resource impacts the livelihood of the people, and hence, conservation should ensure sustained productivity for meeting the present and future needs of the dependent community. Obviously, the community is expected to be actively involved in planning and management of the resource.

Conservation of minerals could be achieved in several ways. One of the ways of conservation is by systematic stock piling of overburden (waste material) and low-grade ores which have no demand under the prevalent marketing conditions. Such ores could be utilized in future, whenever there is demand for these grades of ores. If they are not properly stock piled and allowed to get mixed with the overburden, then they will not be available for utilization in future when the demand for them increases.

Beneficiation of low-grade ores is yet another method by which conservation could be achieved. By beneficiation (or ore-dressing) is meant selective separation of wanted mineral substance from the unwanted material that is gangue. Broadly, physical methods are employed to concentrate the ores; however, at times a combination of both the physical and chemical methods is used. The simplest way by which beneficiation could be achieved is by hand picking the ore from the gangue. In some cases, it could also be done by sieving the ore, provided there is a grain-size difference between the ore mineral and the gangue. Washing the ore is the most commonly employed beneficiation technique, especially when there is sufficient density difference between the ore mineral and the gangue. This technique is widely employed to upgrade the magnetic ores from south Goa. There are 28 beneficiation plants in the State. Availability of large quantities of water is, however, a pre-requisite. Water is often recycled to conserve its usage. Despite precautions, large-scale beneficiation by washing can lead to environmental problems, such as pollution of groundwater aquifers and surface water bodies.

Magnetic separation is yet another technique of beneficiation. The iron ore mines from south Goa employ both of these techniques. The magnetite-rich iron ore is first washed and then subjected to magnetic separation. Silica being the prime impurity most of it is removed during washing, other impurities are removed to a considerable extent by magnetic separation.

Yet another method which is often employed is ***blending*** of ores. The high-grade ores are suitably blended (mixed) with low-grade ores, so that the grade of the mixture is as per the industrial requirements. By employing this method, the consumption of high-grade ores is reduced, and at the same time, the low-grade ores can be put to effective use. In other cases, heavy liquid separation can also be attempted; however, it is a very expensive method which is resorted to only in case of very expensive minerals, that is, those which have a high 'unit value'.

Some of these beneficiation techniques are fairly expensive when manual labour is employed, since in a mining operation, which is on a commercial scale, the quantum of material involved is in thousands and millions of tons. This is especially the case in underdeveloped and developing countries where required technology is not available or is unaffordable; hence, manual labour is the only available recourse. The non-availability of suitable technology for beneficiation could then make a potential ore deposit a non-deposit as the cost involved in employing manual labour is unaffordable. Hence, in the developing world, there is an increasing awareness and demand for acquisition of new technologies and upgradation of the existing ones. This could be employed to beneficiate the low-grade ores at lower costs, thereby conserving the high-grade ores and suitably utilizing the low-grade ores either for indigenous consumption or for export.

3.7.4 The National Mineral Policy

The National Mineral Policy enunciated in 1993 and amended in 2008 and 2019, outlines the directions desired for the development of mining industry. It prioritizes the action plan essential for the development of mineral deposits and for the conservation of the environment. It is an expression of the vision of the country to prioritize areas of development in the concerned discipline/field. It outlines the approach and the methodology that should be adopted to foster growth and development of the country. Therefore, the policy document outlines the principles and the strategy to be followed in the development of mineral exploitation in a way that is beneficial to the society at large, at the same time avoiding wastage of the precious resources and simultaneously ensuring that least damage is caused to the environment as a result of the exploitation. ***The policy document expresses the intent of the government, but at the same time does not specifically lay down any statutory provisions for development***. Based on the policy document, the statutory provisions are worked out, that define the legal framework for the mining and environmental management.

The salient features of the National Mineral Policy enunciated in 1993 and subsequently amended in 2008 are summarized below:

The guiding principle enunciated in the policy document for the development of mineral-based industry shall ordinarily be the economic factor. However, the State may undertake exploitation of any mineral in public interest so as to ensure unhindered supply of raw materials to achieve the national goal (National Mineral Policy 1993; 2008).

Conservation of minerals shall be understood not in the restrictive sense of abstinence from consumption or preservation for use in the distant future, but as a positive activity leading to the extension of utility of the reserve base through improvement in mining methods, beneficiation for utilization of low grade ores and rejects, and recovery of associated minerals. There needs to be an adequate and effective legal and institutional framework to implement zero-waste mining as the ultimate goal and a commitment to avoid sub-optimal and unscientific mining methods. Non-adherence to the mining plan based on these parameters shall attract punitive action on the mining company. Mineral sectoral value addition should be encouraged through latest techniques of beneficiation, calibration, blending, sizing, concentration, pelletization, purification and general customization of product.

Mine development and mineral conservation, as governed by rules and regulations, will be carried out on sound scientific basis, with the regulatory agencies. Conditions of mining leases shall be such as to lead the lease areas to systematic and complete extraction of minerals and ores.

Since mining contributes to the generation of wealth and creation of employment independently, it should be treated as an economic activity in its own right and not as a subsidiary activity of manufacturing industry. The user industry will be encouraged to develop long-term linkages with mineral production units including equity participation with other mining concerns.

Indigenous industry will be strengthened for the manufacture of mining equipment and machinery. Induction of foreign technology and participation with companies abroad will be encouraged. Import of equipment and machinery, which improve the efficiency, productivity and economics of mining operations and safety, and health of the workforce and people in surrounding areas, shall be freely allowed.

So as to improve the competitive edge of the national mining industry, emphasis shall be laid on mechanization, computerization and automation of the existing and new mining units. The manpower development programme shall be suitably reoriented for the purpose. A comprehensive review of the sector's manpower needs will be carried out and the educational activity will be channelized to meeting these requirements in the medium and long term.

Mineral deposits are generally located in remote and backward areas with poor infrastructural facilities, which hinder their optimum development and utilization. A major thrust on linking infrastructure facilities in mineral bearing areas will be undertaken. A suitable environment will be created to motivate large mining companies to undertake construction of transportation networks (road and rail) on their own. A much greater emphasis will be given to development of health, education, drinking water, road and other related facilities and infrastructure in mineralized belts. This would serve to create an integrated approach encompassing mineral development, regional development and the social and economic well-being of the local stakeholders and of the tribal population.

Mining is an activity that requires monetary support from financial institutions. Steps shall be initiated to facilitate financing of mine development and exploration integral to the mining project. Prospecting being a high-risk venture, access to 'risk funds' from capital markets and venture funds will be facilitated. Introduction of foreign technology and foreign participation in exploration and mining for high value and scarce minerals shall be pursued. Foreign equity investment in joint ventures with Indian companies for exploration and mining will be promoted and encouraged.

Efforts will be directed to promote small-scale mining of smaller deposits in a scientific and efficient manner so as to safeguard the environment and protect the ecology of the region. The growth of illegal mining shall be prevented through tightening of regulations on the conditions for mining. Where small deposits are not susceptible to viable mining, a cluster approach will be adopted by granting a single lease to several deposits within a geographically defined area. Efforts would be made to grant such mineral concessions to consortia of several companies of small-scale miners so that such clusters of small deposits will enable them to reap the benefits of economies of scale.

Mining activity often leads to adverse environmental impact such as land degradation in opencast mining, land subsidence in underground mining, deforestation, atmospheric pollution, pollution of rivers and streams, and soil erosion due to disposal of solid wastes such as overburden among others, all of which affects the ecological balance of the area. Prevention and mitigation of adverse environmental effects due to mining and restoration and re-vegetation of the affected green areas in accordance with the internationally acceptable norms and procedures of afforestation shall form an integral part of the mining industry.

All mining shall be undertaken within the parameters of a comprehensive 'sustainable development framework' which will be the guiding principle to take all these aspects into consideration. The code shall be that a miner shall leave the mining area in better ecological shape than it originally was. Mining operations shall not ordinarily be undertaken in ecologically fragile and biologically rich areas. Strip mining in forest should be avoided to the extent possible. Wherever permitted, it should be mandatorily accompanied by a comprehensive, time-bound reclamation programme. No mining lease would be granted to any party, private or public, without a proper mining plan, which shall include the environmental management plan, approved and enforced by the statutory authorities. The environmental management plan should adequately provide for controlling the environmental damage, restoration of mined areas and reforestation according to the prescribed norms. To the extent possible, reclamation and afforestation will proceed concurrently with mineral extraction.

As the mining operations often involve acquisition of land held by individuals including those belonging to the weaker sections, a social impact assessment will be undertaken to identify areas that will ensure minimum impact, especially on the marginalized sections, and ensure that suitable relief and rehabilitation packages are provided to the individuals/groups concerned. Provision of appropriate compensation will form an important aspect of the sustainable development framework. In so far as indigenous (tribal) populations are concerned, the Framework shall incorporate provisions to protect the stakeholder interest in the mining operation, especially in situations where the weaker sections such as the local tribal populations are likely to be deprived of their means of livelihood as a result of the mining activity. An assessment of the economic, environmental and social impact on the affected people will be undertaken. A sustainable income package would be made available to those affected by mining so as to improve their living standard. For this purpose, the provisions of the National Rehabilitation and Resettlement Policy or any revised Policy or Statute that may come into force, will be made applicable and followed.

Subsequent to the completion of the extraction of ore, a scientifically formulated mine closure plan shall be implemented by the company so as to restore the ecology, regenerate the biomass and ameliorate the socio-economic impact on the stakeholders.

The mining operations are hazardous in nature; accidents occur and often result in the loss of life or disability to persons engaged in mining. Efforts shall be directed towards development and adoption of mining methods that would cater to the safety of workers and reduction of accidents. Likewise steps will also be undertaken to minimize the adverse impact of mining on the health of the mine workers and the population in surrounding region.

Thus, it is clear that the National Mineral Policy enunciates the guidelines to enact appropriate legislation to ensure systematic, planned and scientific mining. It makes it obligatory to take care of the natural and social environment so as to avoid pollution, deforestation, degradation and social unrest, and to enforce reclamation and restoration of the environment. ***The policy, therefore, does not* per se *lay down the legislation or does not frame rules to regularize and systematize the exploration or mining of a resource*. *It merely provides the direction to the government of the time to frame adequate rules and regulations to control the exploitation of the natural resource*.** In doing so, there is a check at every stage of exploration and mining which keeps the entire process of mining under control, right from the time of obtaining a mining lease, up to the utilization or export of the mined product. Every operation is dictated by regulation thereby leading to the protection of the environment.

3.7.5 Mines and Minerals Regulation and Development Act

The 'National Mineral Policy' of the Government of India therefore *inter alia* incorporates guidelines/provisions for the formulation of legislation necessary to enact rules and regulations intended for the exploitation of mineral deposits. These include the Mines and Minerals (Regulations and Development) Act (MMRD Act) of 1957 (amended in 1958, 1960, 1972, 1978, 1986, 1994, 1999, 2010 and more recently in 2015), Mineral Concession Rules (MCR) of 1960 amended in 2016 and the Mineral Conservation and Development Rules (MCDR) of 1988 amended in 2017.

The salient features of the MMRD Act of 1957 (with amendments as outlined above) to bring it in conformity with the National Mineral Policy of 1993 amended in 2008 are:

(i) Thirteen minerals have been dereserved for exploitation by the private sector. These include iron ore, manganese ore, chrome ore, sulphur, gold, diamond, copper, lead, zinc, molybdenum, tungsten, nickel and platinum group metals.
(ii) Any company registered in India, irrespective of its foreign equity holding is allowed to apply for a mining lease or for a prospecting license.
(iii) Fifteen minerals are excluded from the first schedule of the MMRD Act. These are apatite and phosphatic ores, barites, dolomite, gypsum, vanadium, kyanite, sillimanite, magnesite, molybdenum, nickel, platinum and other precious metals, silver, sulphur and its ores, tin, tungsten and vanadium ores. Hereafter, state governments are not required to obtain prior approval from the

Central Government for the grant of mining lease or prospecting license for these minerals.

(iv) Prior approval of the Government of India is required for mining leases of eleven minerals apart from atomic minerals and mineral fuels. They are asbestos, bauxite, chrome ores, copper ores, gold, iron ores, lead, limestone (except where the usage is in kilns for the manufacture of lime as building material), manganese ore, precious stones and zinc.

(v) The period for which prospecting license can be granted is increased from two to three years. The license can be renewed up to five years by the state governments.

(vi) All mining leases to be issued for a maximum period of 30 years. The minimum period of lease shall be not less than 20 years. All leases shall be granted for a period of 50 years from the date of commencement of MMRD Act 2015. Those leases granted prior to this date shall be deemed to have been granted for 50 years (b) in order to take care of the areas affected by mining, the State Government is directed to establish a trust called 'District Mineral Foundation' to which the lease holder shall contribute an amount equivalent to 10% of the royalty. For older leases, granted prior to 12 January 2015, the contribution shall be equivalent to 30% of royalty, (c) the Central Government shall also set up a National Mineral Exploration Trust to which 2% of royalty shall be contributed; (d) in the case of notified minerals, the State Government shall grant the lease through auction by competitive bidding, including e-auction (amendment in force from 12 January 2015).

(vii) In case no work is undertaken, the period before which a mining lease can lapse has been increased from one year to two years.

(viii) New section has been introduced allowing searches to be carried out to check unauthorized mining.

(ix) State governments are authorized to terminate mining leases of minor minerals without the prior approval of the Central Government.

(x) No appeal or revision to lie with the Central Government for orders passed by the State Government with regard to the minor minerals.

(xi) Notifications issued by the state governments under minor mineral concession rules to be placed before the State Legislature.

Central Government is empowered to modify leases as per the Act, to bring them in conformity with the provisions of the said Act.

Additional details as regards the operational part of mining of minerals are provided through Mines and Minerals Regulations and Development Act, 1957, and amendments (listed above), Mineral Concession Rules, 1960, 2016, and Mineral Conservation and Development Rules, 1988, 2017. It has been made mandatory for the mining concerns to provide self-appraisal reports on the basis of which they would be rated by competent authorities.

3.8 Mining of Mineral Resources

Modern civilization is primarily the result of two basic industries, namely agriculture and mining. The former provides for sustenance, and the latter caters to the supply of mineral raw materials required for both the infrastructure and industry on which depends the growth of the civilization. Thus, minerals constitute the backbone of economic growth of the nation.

Mining is a commercial activity of extraction of natural wealth from the ground/earth and selling the material/produce at profit (for economic benefit). The economic aspect (profitability) being the prime concerns in mining, obviously it is intimately linked to production and productivity. If production refers to the quantum of ore that is mined, productivity involves all aspects that affect the profitability of the mining venture. During the post-nationalization period, major minerals required by the industry were under the control of public sector undertakings, e.g. Coal India Limited, Hindusthan Copper Limited, Hindusthan Zinc Limited and others which did not have to compete with any other mining organization/company. The funding required for sustenance and development was provided by the government which supported a high cost economy with low productivity. A large workforce and heavy mechanization coupled with low utilization and sluggish production were the hallmarks of the public sector undertakings.

With the liberalization of economy and free marketing in 1991, the public sector mining industry had to witness stiff competition from private players both from India and abroad. It was obvious that the cost of production in public sector undertakings increased due to low productivity. In comparison with the cost of indigenous ores, imported mineral concentrates worked out cheaper causing drastic recession in the public sector mining industry. The immediate changes that were noticed were closing down of several mining establishments and attendant retrenchment of the workforce leading to societal problems primarily due to unemployment.

3.8.1 Mining Activity in Goa

Nearly one-eighth of the area of the State is influenced by mining activity. In addition, ancillary activities associated with mining encompassed still larger areas. The total number of iron ore mining leases (erstwhile mining '***concessions***' of the pre-liberation era) approved for exploitation by the then colonial (Portuguese) regime was 148, covering a total area of 10,839 hectares, which amounted to 2.93% of the total area of the State. Of these 88 lease areas, encompassing an area of 6,661 hectares (i.e. 1.8% of the total area of the State) was under active mining activity as of 2004. Post-mining ban (2012), the available figures of lease areas are at variance in different departments of the government. However, at the time of the imposition of the mining ban, there were 337 valid leases (officially declared) of which about 100 were in operation. There is an implicit difference between the term '*Concessions*' of pre-liberation era and the 'mining leases' of the present times. Under the colonial mining laws, the titles of the 'concessions' were granted to the holders in perpetuity and were recognized as patrimonial rights by the Portuguese government vide the decree dated 20 September 1906, so that the holder could mine the area either till the ore got exhausted or till mining was uneconomical, without requiring to undergo the process of renewal of the mining lease holding, as laid down by the National Mineral Policy of 1993 and MMRD Act of 1957 and its amendments. The mining concession thus, was 'unlimited in duration as long as the concessionaire complied with the conditions, which the law and title of concession imposed on him'. This provision of perpetuity of rights was abolished by the parliament amending the Goa Daman and Diu (Abolition of Concessions and Declaration as Mining Leases) Act in 1987.

In consequence of this amendment, the 'concessions' became mining leases as per the MMRD Act of 1957, amended several times, the last in 2015, and the leases were required to be renewed as per the laid down procedures, in which case there was a possibility that a lease may go to a new lessee during the process of renewal. This is not desired by the private mining companies. The State on its part seemed inclined to tow the line of the mining companies by amending the Goa Daman and Diu (Abolition of Concessions and Declaration as Mining Leases) Act of 1987. Bringing back such harsh provision of perpetuity of concessions which is not in the larger interest of the State, primarily for two reasons, among several others. One, the State is too small, and therefore, the resource base is very limited. The normal strategy of private mining corporations is one, to work the deposit as fast as possible till it is exhausted or is uneconomical and then shift the company's base elsewhere, and two, there is a tendency among some corporates that as soon as the company realizes that the activity is uneconomical, a proposal is moved for the mine closure with a very appealing mine closure plan. There have been instances elsewhere, when such mine closure plans were proposed not due to paucity of the resource or its exhaustion, but the mining activity was uneconomical due to some other factors. In such closures, if permitted, the remaining resource is lost for the State forever and this is not in keeping with sustainable mining policy. Therefore, various ways of conservation of resource and efficient management need be adopted specially by a small state to extend the life of the resource.

Excavation for the recovery of iron, manganese, bauxite and other earth materials required by the construction industry involves removal of vegetation, the soil cover and the overburden. For every ton of ore mined, the minimum overburden generated is nearly twice to thrice that quantity. This is generally stacked in the vicinity of the mine (Fig. 3.2). Excavation and stacking of overburden lead to the modification of the topography of the region that has considerable influence on the circulation of surface water and variation of groundwater table. A much larger area surrounding the mining belt is stressed because of the construction of road network, residential colonies and ancillary industries which have a multiplier effect on land degradation.

The areas stressed by mining activity are likely to increase the world over, specially so in the developing/third world countries such as India. Unfortunately, Goa has already scaled the peak of environmental damage due to mining. There can be

Fig. 3.2 Large, extensive dumps of overburden in North Goa

perhaps no further escalation in degradation due to this activity, as nearly 80–90% of the recoverable reserves have already been mined, unless new deposits hitherto not reported are identified for exploitation. The learned bench of the Supreme Court while banning mining in Goa had observed: '*rapacious and rampant exploration of natural resources is the hallmark of iron ore mining sector-coupled with a total lack of concern for the environment and the health and well being of the denizens in the vicinity of the mines. The sole motive of mining lease holders seems to be to make profit (no matter how) and the attitude seems to be that if the rule of law is required to be put on the back burner, so be it*'.

3.8.2 Relationship of Type of Mining and Extent of Damage

Geology and the geomorphology of the terrain primarily decide the type of mining to be adopted. Physical properties of the rocks such as cohesivity and competence / strength of rocks are of prime concern in deciding the mining method. Broadly, two types of mining operations are carried out: (i) open cast mining and (ii) underground mining. The first type is generally undertaken where the enclosing rocks are less competent and soft, the ore body is at a relatively shallow depth and the terrain is less rugged. It could be done in strips (strip mining) or along contour lines in mountainous areas (contour strip mining). Underground mining is resorted to where the enclosing rocks are hard and competent (generally unfractured) and the ore body is deep seated. The ruggedness of the terrain is also a factor to be considered.

Opencast mining is undertaken for the exploitation of all the three minerals, namely iron, manganese and bauxite in the State. Contour strip mining is the commonly followed method. This involves removal of overburden beginning from the outcrop or from the summit of the hillock. Gradually, the quarrying proceeds backwards and downwards. As the height of the working quarry-face increases progressively, it is so excavated that the mining pit has a step-like appearance, which in the technical jargon is referred to as contour strip mining (Fig. 3.3). Each step is referred to as a bench. Height of each bench (i.e. each working face) is maintained as per stipulation (mining rules) to be not more than 8–10 m.

In large mines, worked by major mining concerns, especially in the iron ore industry, a systematic mine development is undertaken. The rock being mined is soft and it is mechanically quarried. It is easier to follow contour strip mining. However, in the case of mining of hard rocks for the road metal, the quarries do not conform to mining regulations, and no benches are maintained since it is not cost effective, difficult to quarry and requires repeated blasting with explosives.

Fig. 3.3 An open cast iron ore mine in North Goa depicting contour strip mining

3.8.3 Slope Stability

In all mining operations whether large or small, slope stability is a matter of great concern, as it affects the safety and security of man and machines.

Slope refers to the ground slope which may be an inclined surface, either natural or man-made, which is at an angle with the horizontal. Slopes related to mining consist of number of different parts called ***facets***, if straight/rectilinear, and ***elements***, if they are rounded—each element having its own inclination. In order to fetch precious minerals, the material has to be excavated from the ground either manually or mechanically due to which an excavation/pit is created.

The unwanted material known as the '***overburden***' is dumped separately and the precious mineral is stacked separately. During this process, the excavation gets deeper and is bound by inclined surfaces which get continuously modified as mining proceeds. The angle of the inclined surfaces needs to be maintained below a critical angle, so that the material on the surface does not move down or cave-in under the action of gravity. This is known as maintaining the slope stability. In addition to this, sloping surfaces are also generated due to the storage of ore and also stacking of the overburden. All these and similar inclined surfaces which are formed as a result of mining activity are included under the purview of ***slopes*** and the maintenance of the angle of repose goes under the head of ***slope stability***.

It is a common sight to find slopes of hills covered by loose material that is prone to some form of movement downslope under the action of gravity. This type of movement which leads to erosion is known as ***Mass Wasting*** and the actual downslope movement itself is known as ***Mass Movement***. It may include slow creep of clay-rich soil on gentle hill slopes or the sudden and rapid fall of large boulders of rocks from hanging cliffs. All such phenomena can be broadly included under ***slope failure***. It forms a part of the process of landform development during which material constituting a slope adjusts itself to the surface angle. It is a response of the slope material to the changes in hydrologic, climatic, geomorphic and biotic factors. However, the term 'slope failure' more specifically by convention is used in relation to failures of slope in mining areas.

Mass movement is viewed as a potential hazard more because of its catastrophic effects especially in populated areas than its erosional effect of removal of soil from the hill slopes. Although mass movement is a natural phenomenon, it is accelerated by anthropogenic/human activity. The hazards of mass movement are increased as human population increases. This leads to faster urbanization of newer areas where settlements develop without adequate care for safety and precaution to prevent an eventuality due to mass movement.

3.8.4 *Factors Responsible for Slope Movement*

All materials are acted upon by gravitational forces. Rock and soil on sloping surfaces are not exception to it. Sometimes, in addition to gravitational forces tectonic forces are also active. Rocks and soil, i.e. the mass, as long as it is able to withstand the forces that are acting on it, remain stable; that is, it remains in equilibrium. However, any imbalance in forces brought about either by natural phenomenon or by anthropogenic activity may result in the movement of the mass downslope, which is generally referred to as slope failure. It may be sudden and short-lived or may be slow and long-lived. Failures may occur due to the man-made activity such as in mines and other construction activities such as buildings, dams, bridges and tunnels among others. Failures may also take place on hill and valley slopes and scarp surfaces by the activity of natural agencies such as water, wind and tectonic activity resulting in earthquakes and tsunamis. Some of these may be catastrophic, and hence, it is essential that they are carefully studied to find ways and means by which the risk can be minimized and at times avoided.

3.8.5 *Role of Water in Mass Movement*

Mass movement along a slope occurs where the force of gravity exceeds the resistance of the weathered material (to move). Although the force of gravity does not change substantially from place to place, the resistance of the material does vary. It varies both with place and time. Its variation in space depends on its location. On steeper slopes, mass movement is facilitated whereas gentler slopes usually do not support mass movement unless they are tempered with by humans. One good example of mass movement caused due to the human interference with nature was observed during the construction of rail route (Konkan Railway) from Mumbai to Trivandrum along the western coastal belt of India. In the eagerness of completion of the project within the minimum stipulated time frame, unplanned and at times unsystematic hill-cuts were undertaken without taking adequate precaution for the slope stability. Whereas, the normal slope angle for stable slopes is less than 30°, the cut sections of the railway tract had slopes as high as 50°. The oversteepening and the undercutting of the slopes greatly reduced the safety factor. Several landslides and rock-falls thus occurred, especially in the rainy season, causing loss of life and property.

The variation in time is related to the accumulation of weathered debris which over a long time period may exceed the force of resistance due to the constant accumulation and may therefore slide downslope. Such movements are generally controlled by the quantity of water in the debris. The force of gravity acting on a sloping surface has two components. One acts perpendicular to the direction of the slope and helps hold the debris in place. It provides frictional resistance against mass movement. The other component that acts in a direction parallel to the direction of the slope serves as the driving force to move the mass downslope. When the slope is steeper, the slope-parallel component of gravity is greater than the slope-perpendicular component, and the slope becomes unstable at some critical angle from the horizontal and failure occurs. The angle above which slope failure occurs is known as the ***angle of repose*** which is about 30–35° for loose dry material such as gravel and sand.

The other factor that is responsible for slope failure is the quantity of moisture in the debris which influences the strength of the material. When the spaces between particles are occupied by water, the strength of the material changes. A small quantity of water provides greater resistance to movement due to surface tension which holds the particles together. Just as loose dry sand and silt role down a sloping surface of an inclined plank, but when made moist they adhere to the plane and hence come to rest on the sloping surface of the plank. However, when the pores are completely filled with water, it exerts a pressure on the grains and pushes them apart leading to slope failure.

Thus, slope failure usually takes place in nature after heavy rainfall of several days. Such slope failures and slides are common in the Himalayas where there are alternations of sandstones and clays that have dips (inclination) of the bedding planes towards the valley floor. Such slopes are called as dip slopes. In rainy season due to excessive precipitation and in dry season due to melting of ice, the clays get saturated with water, and their resistance to movement is considerably reduced. They become plastic and slippage occurs leading to major hazards that involve devastating effects on life and property downslope. Mass movement is less likely, if the beds are horizontal or they dip into the slope, i.e. along the escarpment slope.

3.8.6 Slope Failure

Slope failure is a part of the process of landform development during which material constituting a slope, adjusts itself to the surface angle. It is a response of the slope material to the changes in hydrologic, climatic, geomorphic and biotic factors. In the foregoing, the aspects of slope failure due to mining activity are solely considered.

Slope failure affects the environment in some form or other. The effects may be either visible immediately, readily perceptible, large and catastrophic or they may not be perceptible, small and slow in having a significant effect on the environment. The changes that occur may or may not have an effect on human life and the society, depending upon the surroundings and the ecology of the site in question.

Slope failure is related to the extent and the frequency of mass movement that may create problems to the human activity or for the land use. Slope failure is often seen in the context of social relevance and involves threats to the slope stability in relation to the life and property. Inherent in this concept is the recognition of *hazards* and *risks* in land-use planning.

Hazard here refers to the slope failure in terms of its potential magnitude. A careful study is capable of predicting a potential hazard. In other words, the study predicts the slope failure with specific reference to the area of influence, volume of the mass involved and the velocity at the time of impact. Hazards may be categorized as of low, moderate, high and very high intensity. Those classified as 'very high' normally involve potential loss of human life.

Risk relates to the consequence of the slope failure on the land-use activity or human activity, for example, inconvenience or damage and/or destruction of civil structures such as buildings, roads, railways, bridges and/or loss of life.

3.8.7 Impact of Slope Failure

Slope failure may lead to hazards of a low degree, but the risk involved may be very high. This is commonly encountered in slopes created by surface excavations, especially in opencast mines where a rock- or a boulder-fall in a working face may cause severe loss of life and damage to the machinery. Thus, slope failure in working mines adversely affects both the functioning and the economics. Similarly, the hazards of low to moderate degree may also have serious socio-economic impact particularly in the case of settlements such as villages, towns and cities situated on the hill slopes. Most hill stations, particularly those located in the Himalaya such as Kasauli, Kullu, Manali, Shimla, Darjeeling among others, are prone to hazards of slope failure. A low degree hazard in such cases may involve very high risk of loss of life and property. Similar is the case of slope failure involved in civil constructions such as the dams, canals, bridges, hydel power stations and hilly roads, where slope failure could lead to severe loss of life, apart from damage to the property.

Slope failure resulting from human (anthropogenic) activity is, however, less catastrophic than some hazards triggered by natural processes. For example, in the year 1920, an earthquake-triggered loess flow in China put to death an estimated 20,000 people (Close and McCornick 1922). Similarly in 1962, a debris avalanche from the north peak of Nevados Huascaran in the Peruvian Andes took a toll of 4,000 lives (Rouse 1984).

3.8.8 Type of Mass Movement

Mass movement can be categorized under several heads, depending upon the type of debris involved, the amount of water and the speed at which the debris moves. Three basic mechanisms of mass movement are identified: ***heaves***, ***slides*** and ***flows***, and are distinct from the pure free fall of rock debris that are known as rock-falls. ***Heave*** is caused by alternating expansion and contraction of debris due to freezing and thawing or wetting and drying that tends to raise and lower the debris in a direction perpendicular to the slope. ***Slides*** take place when the cohesive rocks fail along well-defined slip-planes such as in the case of the alternating clays and sandstones in Himalaya, mentioned previously. ***Flows*** are generated when the debris move more like a fluid with no specific plane of failure, as in the case of a slide.

The most common type of mass movement that can be attributed to heave is ***soil creep*** which is a slow movement that occurs in the presence of substantial amount of water. The tilted fence posts and power poles on slopes are usually the result of soil creep.

Rapid mass movement such as rock slides, debris slumps and debris slides occur essentially without the involvement of water. In cases where the involvement of water responsible for mass movement is more, the processes are identified as debris flows and mud flows. In the latter case, the percentage of water relative to the debris is much higher as compared to the debris flows. Debris flow may contain about 10% water, and the flow is, therefore, very viscous and is capable of transporting huge boulders and even buildings. Mud flows, on the contrary, are more fluid and contain about 30% water.

***Triggering Mechanism*:** The two factors responsible for mass movement are internal and external. Both of which act in unison over a long period of time to reduce the slope stability. However, sudden and abrupt failure may be triggered by heavy rains, earthquake shocks, shocks from volcanic eruptions, rapid withdrawal from a reservoir or large-scale excavation.

3.8.9 Management of Slope Stability

The management of slope stability is primarily governed by the economic considerations and the risk involved, which could be gauged from the hazard prediction. Therefore, one of the prime requirements in the management of unstable natural slopes such as those in mountainous terrains is the preparation of hazard zonation maps. The Himalayan region has been mapped for hazard zonation by several workers (e.g. Anbalangan 1992; Kishor Kumar et al. 1996). Despite such studies, the loss due to slope failure by landslides is quite high as the terrain is characterized by unstable natural slopes. Failures are quite frequent both spatially and temporally. The loss due to such repeated slope failures may be much higher than any single natural hazard. This is frequently noticed during the construction of roads in the Himalaya. The cost involved at the time of completion of the project is often much higher than the cost estimated at the time of initiation of the project. This increase in cost is not entirely due to escalation because of inflation, but is due to escalation in cost required for clearance and reconstruction of structures damaged by landslides during the course of construction.

The hazardous and unplanned excavation often leads to slope failure causing damage to human life and property (e.g. Highland and Brown 1996). In order to avoid risk to life and property, structures on slopes are designed in such a way that they are within an empirically decided safety limit which is known as ***safety factor***. The latter depends not only on the probability and the magnitude of hazard but also on the risk factor, which in fact is the most important controlling factor in the design of engineering structures. Choice of safety factor varies depending on the expert group. In general, safety factor ranges from 1.0 to 1.5 depending upon the predicted hazard, risk involved and the slope life.

Artificial slopes formed by dumping of overburden, industrial waste and garbage have been a cause of concern in recent times. Mine dumps have attracted more attention due to the quantum of material involved and the fast rate at which the material is stacked, so much so that there is no time for the material to settle itself. The slope stability is the concern of the safety of mine workers, the machinery that works on the dump platforms and the adverse effect on the environment due to potential slope failure. As the quantum of material involved is large and the growth of the dump is rapid, the stability of the dump is greatly influenced, increasing the potential/chances for failure. Slope failure may be fatal to the mine workers. It may also lead to loss and damage of equipment, cause blockage to the public or private access, damage to public and private land, depending upon the dump site location, topography and overall social relevance of the location. Slope failures have been quite common in opencast mine dump sites of Goa particularly during the pre-liberation times when adequate precautionary measures were not taken with regard to the overburden dumps.

In comparison with mine dumps, the fatalities due to slope failure in case of garbage dumps are relatively less, since except in few cases, the dumps are less extensive and the workforce involved is relatively smaller, in addition to the garbage pickers. Moreover, to the extent possible, the dump site selection is done with adequate regard to the location and topography of the terrain, so that the adverse effect on the nearby settlement is minimal.

In all these cases and particularly in the case of mine dumps, slope failure primarily affects the downstream areas. This may occur due to the sliding of loose material from dump slopes, particularly when charged with water in the rainy season, affecting structures downstream. Similarly, fine particulate matter such as clay and silt may be carried and spread over agricultural land impairing its fertility and often rendering it uncultivable. Transport of loose material by flowing water also affects the natural surface flow in general, in addition to adversely affecting aquatic life and ground- and surface water quality.

3.8.10 Preventive Measures

As mentioned earlier, mass movement primarily occurs due to either of two forces that act on the mass. One is the increase in slope component of the force of gravity and the other, the decrease in the force of resistance or the reduction in the strength of the mass. Ordinarily, the increase in gravity component and the decrease in the strength of the mass usually take place when naturally occurring slopes are cut and steepened for various purposes such as for the construction of means of communication such as roads, railways, canals, dams and bridges among others. The excavation of the already stabilized material from the lower levels results in loss of support for the material above. It then loses the resisting force along a potential plane of failure, resulting in a slide. So to prevent mass movement either the gravitational component has to be reduced or the strength of the material got to be increased. Prevention of failures requires a judicious alignment of roads, railways and other cuttings in a manner that will least destabilize a slope. Additionally, construction of retaining walls, benches, terraces and more recent techniques such as the use of geotextiles can minimize slope failure.

Natural phenomenon can also lead to slope failure; for example, under-cutting of cliffs by wave action or stream currents can reduce the support of the rock at the base of the cliff, leading to mass movement. Similarly, reduction in resistance or diminution of the strength of the material could also take place by addition of water to the slope debris. This could happen especially along newly constructed roads, railways (an example cited above), canals and leakages from sewage and water pipes.

In all such cases, the only mitigation measure is to prevent water accumulation and allow channel ways for its escape. Vegetation can also play a significant role in controlling mass movement. Roots of plants hold the material together and provide strength to the hill-slope debris. Hence, deforestation and removal of plants and shrubs from hill slopes should be prevented. Vegetation also absorbs moisture and thereby helps reduce the accumulation of water in the soils on slopes. One of the ways of mitigation of mass movement on deforested and devegetated slopes is to undertake afforestation and mulching measures for the new plants to take roots. This has been made compulsory in the mining belts in India. In certain areas, mass movement has been a common phenomenon in the past due to haphazard mining methods employed to save on costs. This in many cases has led to loss of life, even in Goa where mining is much more systematic as compared to the rest of the country.

It is essential that a geoscientist maps the area to know the type of soil and debris, its characteristics, rocks formations beneath the debris, incidents of earlier mass movement and potentially hazardous areas. Such slope-hazard maps can then be used for a systematic planning, strengthening of engineering structures, stabilizing slopes and preventing their failure.

3.9 Opencast Mining

One of the advantages of opencast mining is, it is more in tune with the mineral conservation policy which states ‘maximum utilization of resource avoiding wastage’. This can be fulfilled to a large extent as greater recovery of ore can be achieved in comparison with that in underground operations. Air pollution is relatively less at the actual site of mining but is quite significant during the transport of ore and overburden. Most road network in the State being of laterite aggregate, lot of dust-pollution takes place due to the movement of vehicles. In terms of safety of the workforce, open cast mining is safer and the risks/hazards involved are much less as compared to underground mining, provided mining is carried out as per the stipulated norms and observing safety precautions.

3.9.1 Limitations of Open Cast Mining

The risks involved in this type of mining include cave-ins of mine faces (quarry walls) particularly when adequate safety measures are not followed. One of the worst disasters in recent times occurred on 9 December 2007 at Tudou mine in South Goa that involved loss of several lives and machinery worth over ₹60 million (*Navhind Times*, dtd. 10.12.2007). Several mine workers/ labourers were drowned in the mine pit filled with water. In fact, they were buried alive due to the collapse of the working face, where work was in progress. Similar accidents have occurred in the past in several mines in Goa, particularly during pre-liberation period when enforcement of mining laws was less stringent. In recent times, uncontrolled and unregulated mining activity is more prevalent in South Goa where it is essential to strip away large quantity of overburden to reach the ore zone (pay zone). During the initial boom period, the mines were sub-let on contract to small firms involved in mining, which to save on costs resorted to mining methods with utter disregard to stipulated norms and procedures.

References

Anbalangan R (1992) Landslide hazard zonation and evaluation mapping in mountainous terrain. Eng Geol 32:269–277

Brundtland Report (1987) Our common future, United Nations, http://www.are.admin.ch

Close U, McCornick E (1922) Where the mountains walked. An account of the recent earthquake in Kansu province. China. Natl Geogr Mag 41:445–464

Cohen JE (1995) How many people can the Earth support? W. W. Norton and Company, New York, 532

Highland LM, Brown WL III (1996) Landslides-the natural hazard sleepers. Geotimes (January):16–19

Keller EJ (1976) Environmental Geology. Charles E. Merrill Pub. Co., Ohio, p 548

Kishor Kumar, Tolia DS, Satish Kumar (1996) Landslide hazard evaluation in part of Himalaya. Proc. 7th Inter. Symp. on Landslides, Trondheim, Norway, pp 17–22

Merritts D, De Wet A, Menking K (1998) Environmental geology. W. H. Freeman & Co., New York, p 452

Mines and Minerals (Development and Regulation) Act, (1957), mines.gov.in

Mines and Minerals (Development and Regulation) Act, Amendment Bill (2008), prsindia.org

National Mineral Policy (1993) Mining Policy and Legislation, mines.gov.in

National Mineral Policy (2008) Ministry of Mines, Government of India, mines.gov.in

National Mineral Policy (2019) Ministry of Mines, pib.gov.in

Munn RE (1992) Towards sustainable development. Atmos Environ A Gen Top 26:2725–2731

Rouse WC (1984) Flow slides. In: Brunden D, Prior DB (eds) Slope instability. Wiley, New York, pp 491–522

Tank RW (1983) Environmental geology. Oxford University Press, New York, p 549

Chapter 4
Water Resources

Abstract The erratic distribution pattern of precipitation, demographic pressures and increased anthropogenic activity have taken a toll of the hydrological environments of Goa. The declining trend and degradation of groundwater is a cause of serious concern that needs urgent attention.

Three lithological types characterize the groundwater aquifers in the State, namely (i) the lithomarges beneath the laterite duricrust, (ii) sandy alluvial-colluvial deposits of the coastal plains that abut against the lithomargic aquifers and (iii) weathered and fractured country-rock metasediments/granitic gneisses. The first is an unconfined aquifer, the second is a semi-confined aquifer whereas the third is a confined aquifer. The presence of intercalated clays/lithomarge horizons, at places, gives rise to sub-artesian conditions.

The annual precipitation in Goa varies from 3000 to 3800 mm. The water potential of the State is placed at 85.70 Mm^3 of which nearly 40% goes as runoff. Irrigation accounts for 146.50 Mm^3 of which 112.50 Mm^3 is contributed by surface water and 34.00 Mm^3 from groundwater. The water requirement of the State is 1329 Mm^3. In comparison with the water level of the previous decade (2005–2015), a declining trend of watertable is noticed in the northern, central and western parts of the State. The eastern *talukas* namely Bicholim, Sanguem and Valpoi, which host the iron ore mining belt, exhibit recession in watertable due to dewatering of the mines. Positive impact on the groundwater table is also observed in some parts of these *talukas* as a result of secondary recharge due to the return flow. Increased anthropogenic activity has had an impact on the groundwater quality whereas excessive withdrawal has resulted in saltwater intrusion in coastal areas.

The chapter reviews the experiences and strategies of efficient utilization of water resources and recommends actions for mitigation to control overconsumption of the resource from increased demographic pressures. Efficient ways of rainwater harnessing and water conservation initiatives are discussed to relieve pressure on the groundwater resources and to keep pace with the increasing demand for ensuring sustainability of this fundamental component of the economy.

Keywords Surface water · Groundwater · Aquifer · Salt water intrusion · Rain water harvesting · Artificial recharge

A. G. Dessai, *Environment, Resources and Sustainable Tourism*,
Advances in Geographical and Environmental Sciences,
https://doi.org/10.1007/978-981-99-1843-0_4

4.1 Introduction

Water is the most common substance on earth that is essential for the sustenance of life. Hence, it is known as the elixir of life. It occupies over 70% of the planet and is one of the few substances that can exist in all three states, namely gaseous, liquid and solid. The change in state between and among phases involves the transfer of energy leading to its movement around the globe, from the equatorial regions to the poles. This leads to the global weather patterns. Water, therefore, acts as a climate ameliorator through the energy absorbed and released during its transformation from one state to another. It is, therefore, possible for the earth to have a climate that is habitable for life forms. In modern times with 1.1 °C of warming, climate change is affecting weather and temperature patterns, directly impacting animal and plant life.

Although agriculture by far is a major consumer of fresh water globally, followed by the industry, tourism is not lagging far behind. Tourism and leisure activities can also be the dominant sectors for freshwater use including, swimming pools, irrigation of gardens, golf courses, saunas or spas, in addition to kitchens, laundry, showers, toilets and many other activities. Indirect tourism-related water consumption is predominant in the construction activities. Likewise, biofuel production, which is gaining momentum in recent times, is an water-intensive enterprise. Water is also instrumental in poverty alleviation as it plays a crucial role in ensuring livelihood security in the developing world. It is for these reasons considerable importance is attached to the study of water which is a unique and precious commodity that needs to be valued, conserved, safeguarded and efficiently managed for ensuring the sustainability of the economy.

4.2 Climate and Weather Conditions

India, in general, has a diversity of climatic conditions and a greater variety of weather conditions. The climatic conditions vary from extremely hot in the northwest to intensely cold in the north, with regions of extreme aridity and negligible rainfall to those of excessive humidity and torrential rains. The climatic conditions greatly influence the water potential of the country. The climate over India is primarily influenced by two major physiographic features, the great mountain ranges of Himalaya in the north and the Indian Ocean in the south. The Himalaya acts as the key barrier for the cold winds from central Asia giving the subcontinent a tropical type of climate, while the ocean supplies moisture-laden winds providing a maritime climate. The precipitation in the southern part of the country is largely governed by the Southwest and Northeast monsoons. A major part of the rainfall over peninsular India occurs between June and September due to the influence of the Southwest monsoons, except in the southeastern state of Tamil Nadu where it is contributed by the Northeast monsoons, during October–November. The average rainfall in India is about 1215 mm with considerable spatial variation ranging from less than 100 mm in the northwest, in the state of Rajasthan, to more than 2500 mm in the north-eastern states of Manipur, Mizoram, Tripura, among others.

A large number of rivers drain the 50–80 km wide western stretch of the country (1600 km in length) comprising the states of Maharashtra, Goa, Karnataka and Kerala. All along the length (N-S) of this coastal belt the Western Ghats form a major water divide between the west- and east-flowing drainage of the southern Indian peninsula. It is an asymmetrical divide, since the topography to the west is deeply dissected with steep slopes and swift-flowing rivers with irregular gradients and V-shaped valleys whereas the topography to the east is gently sloping, with meandering rivers that exhibit large deltas at their confluence. The west-flowing rivers are smaller in length compared to those to the east of the scarp. They drain merely 3% of the land area, but carry about 11% of the water.

The climate, weather conditions and the drainage of Goa are described under Sect. 1.5.

4.3 The Science of Water

The science of water is called hydrology (hydro from Greek *hudor* meaning water, and logy from Latin *logia* meaning science or study of). However, contemporary hydrology does not deal with all the properties of water. Modern hydrology is concerned with the distribution of water on the surface of the earth and its movement over and beneath the surface and through the atmosphere.

Although, as per the definition, all kinds of water are included within the ambit of hydrology, here we deal with the study of fresh water that is the primary concern. The science of occurrence, distribution and movement of water below the surface of the Earth is included under Groundwater hydrology or Geohydrology with emphasis on the properties of water whereas the term Hydrogeology differs slightly, in that it lays stress on the geological aspects of the study of groundwater. The saline water is largely dealt with under oceanography.

The focus of this chapter is primarily concerned with providing the basic geological framework on the nature, type and occurrence of the groundwater aquifers and focuses on the utilization and conservation of water resources of the State. Hence, before proceeding further, it is proposed to briefly explain a few relevant concepts in groundwater hydrology that would facilitate understanding the discussion that follows. The following terminology is, therefore, elucidated as an aid in hydrological investigations of the State. For more details on the basic information on hydrology, the reader is referred to De Wiest (1965), Todd (1980), Todd and Mays (2005) among other texts.

4.3.1 Concepts in Hydrology

Water, by itself, is never static, it is always on the move. Yet, there are several areas where it can be considered to be stored, including soil moisture, groundwater, snow and ice, and lakes and reservoirs.

Watershed: Hydrologically, a *watershed* can be defined as an area from which the runoff drains through a particular point on the drainage system (e.g. Tideman 1996). It includes the land and water area which contributes runoff to the common drainage point. The term 'watershed' strictly refers to a divide separating one drainage basin from another. However, in common parlance the term signifies ***a drainage basin or a catchment area***. Thus, each order of a stream has a corresponding watershed. Just as the lower order streams join to form the higher order stream, in the same way the lower order watersheds join and form the watershed of the higher order stream. Thus, combined watersheds of several tributary streams of a major river form a major river basin that drains millions of square kilometres of land.

Runoff may be defined as that portion of the precipitation that makes its way to the streams, lakes or seas as surface or subsurface flow. The runoff depends on several factors such as the topography, lithology and structure, soil and its hydrologic conditions, intensity and distribution of rainfall over the watershed, vegetation cover and also on the anthropogenic activity, primarily the conservation practices followed in the area. The water that infiltrates is stored in the openings and voids within the soil and rocks. The storage of water is vertically distributed in two main zones close to the ground surface, described later. An estimate of infiltration and surface runoff is essential in planning the water resource management of an area.

Aquifer: A unit of rock or unconsolidated sediment has the ability to store, transmit and yield water in usable quantities. It is an underground water-bearing permeable rock that can yield an exploitable quantity of water (Fig. 4.1). In more common parlance, an aquifer could be referred to as a body of water-saturated rock through which water finds its way into wells and springs.

Aquitard: An impermeable geological formation which can neither store nor transmit water. It is a rock with very low porosity that permits limited transmission of groundwater. A solid impermeable rock below or above the aquifer is known as '***aquiclude***' (or *aquifuge*). It is a rock which can store some water but cannot transmit it readily. It is a rock with porosity that is so low that it is virtually impermeable to groundwater. The presence of an aquiclude generates additional hydrostatic pressure within the aquifer leading to the formation of a confined aquifer. If a well is drilled into a confined aquifer, water from the aquifer can rise above its level in the aquifer due to additional hydrostatic (fluid) pressure. This is known as the ***Artesian*** condition.

Broadly, two main types of aquifers exist: (i) unconfined aquifer and (ii) confined aquifer.

Unconfined Aquifer: An unconfined aquifer is one in which the watertable forms the upper surface of the zone of saturation. The watertable is an imaginary surface joining the points of equal hydraulic head over the aquifer surface. It can be described

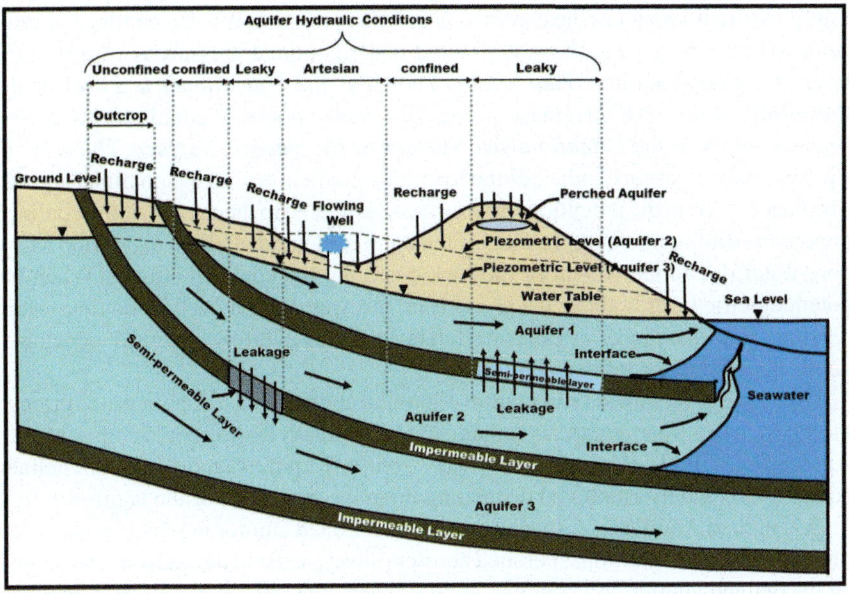

Fig. 4.1 The aquifer types and the groundwater-seawater interface (modified from Wikipedia.org)

as the surface of saturation in an aquifer. It is also called the '***phreatic surface***'. The water below the phreatic surface is referred to as water in the ***phreatic zone***. The water above the phreatic surface or that above the watertable is commonly referred to as '***soil water***'. In other words, soil water is the water in the unsaturated zone, also known as the ***zone of aeration*** or ***vadose zone***. Broadly, water from both the unsaturated and saturated zones contribute to groundwater, however, strictly speaking, water in the saturated zone is referred to as *groundwater* which occurs beneath the '*watertable*'. The watertable slopes in the direction of the hydraulic gradient. The volumetric change in an unconfined aquifer storage is indicated by the rise and fall of the watertable. The thickness from the watertable to the impermeable formation below, which indicates the bottom of the unconfined aquifer, is the saturated thickness of the unconfined aquifer. The unconfined aquifer generally occurs close to the ground surface. When water is able to flow directly from the surface to the saturated zone of an aquifer, it is said to be unconfined. An unconfined aquifer can also be described as a rock unit with a free surface at the upper boundary, known as the watertable and a confining layer at the lower boundary. At the watertable, water is at atmospheric pressure, hence unconfined aquifers are also referred to as ***watertable aquifers*** or ***phreatic aquifers***.

Confined Aquifer: An aquifer that is sandwitched between two aquicludes or confining layers is referred to as a confined aquifer. The confining layers possess extremely low permeability and prevent or restrict the upward or downward movement of water from the aquifer. The water stored in the pores within the confined aquifer is under pressure due to the weight of the rocks overlying the confined aquifer.

This pressure is known as the confined or artesian pressure. When a confined aquifer is tapped by a bore well, the water stored under confined pressure rises above the level of confined aquifer. This level is known as the ***Potentiometric Level*** or the ***Piezometric Level***. An imaginary surface joining the points of equal potentiometric levels is known as the ***Potentiometric Surface* or *Piezometric Surface***. The slope of a potentiometric surface indicates the hydraulic gradient within the confined aquifer. The fluctuations in the potentiometric surface are related to the release of the confined pressure resulting into compression of the aquifer material and the expansion of the pore-water due to the weight of the rock overlying the confined aquifer. When the potentiometric level is above the ground surface, water begins to flow on the surface resulting into flowing or artesian well. The confined aquifers generally develop at depth, below the unconfined aquifers.

If an aquifer (confined or unconfined) has the ability to lose or gain water through the adjacent semi-permeable layer, it is called as a 'Leaky aquifer' or 'Semi-confined' aquifer (Fig. 4.1). The terms such as 'leaky, confined aquifers' and 'leaky, unconfined aquifers' are widely employed depending upon the properties of the aquifers.

A 'Perched Aquifer' is a special type of unconfined aquifer in which water exists under watertable conditions. Perched aquifers always exist in the vadose zone, above an unconfined aquifer or a confined aquifer when a low-permeability layer prevents the downward movement of the water above it. They generally have limited aerial extent and hence are not the targets for groundwater exploration.

The water-yielding mechanism of aquifers differs depending on the type of the aquifer. In an unconfined aquifer, air replaces water in the dewatered zone, as the watertable recedes downwards. Hence, the water from an unconfined aquifer is primarily obtained as a result of the dewatering of pores. On the other hand, in the confined aquifers the pores remain completely saturated and water is not set free due to the drainage of pores. In this case, water is delivered as a result of changes in pore volume due to the compressibility of the aquifer and that of water (change in water density associated with a change in pore-water pressure). Thus, confined aquifers release water due to expansion of water (decrease in water density due to the decrease in pore-water pressure) and compression of aquifer material. Hence, the capacity of confined aquifers to release water from storage is unlike that of the unconfined aquifers.

4.3.2 Properties of Aquifer

The hydrologic factors which govern the storage capacity and the fluid-transmitting ability of an aquifer system are called aquifer properties or aquifer parameters. An aquifer performs two functions, namely storing the water and transmitting it within the aquifer. These two functions are explained by the aquifer properties, namely the 'Storativity' and the 'Transmissivity'.

'Storativity or the Storage Coefficient' of an aquifer is defined as the volume of water taken into or released from the confined aquifer solely due to compression of the aquifer material and expansion of water from the pores per unit change in head over the aquifer surface area. It is a dimensionless parameter. The hydraulic head indicates the 'piezometric level' for the confined aquifers and the 'watertable' for the unconfined aquifers.

The storage of groundwater within the zone of saturation depends upon the hydrological properties of the rock. The storage-related properties include the porosity, effective porosity, specific retention, specific yield, storage coefficient and specific storage. The transmission-related properties, also referred to as yield parameters, consist of the intrinsic permeability, hydraulic conductivity and transmissivity.

Porosity is the percentage of void spaces in a rock. The property of a rock to contain voids, interstices within a geological formation can be referred to as the porosity. The porosity of a porous medium is defined as the ratio of the volume of voids in a porous medium to the total volume of the porous medium. It is a dimensionless parameter. The geological characteristics of the rock formation control the formation of voids and consequently the porosity. For example, size and shape of the grains control the porosity. The smaller grain size and rounded nature of grains give the rock higher porosity. The consolidated sedimentary, igneous and metamorphic rocks have lower porosity. In general, rocks have lower porosities than soils and unconsolidated sediments. Sands, silts and gravel which are made up of angular and rounded particles have lower porosities than soils made up of platy minerals. Similarly, poorly sorted sediments have lower porosities than well-sorted sediments.

'Effective porosity' is defined as the portion of interconnected void spaces in a medium through which a fluid (liquid or gaseous) can flow. A small portion of the total porosity of a medium constitutes the effective porosity that is permeable and is available for the fluid flow. It is also referred to as the 'kinematic porosity'.

'Specific Retention' is the quantity of water held back within the pores under the influence of cohesive and adhesive forces. It is defined as the ratio of the volume of water retained after saturation against the gravity to its own volume. Specific retention increases with decreasing grain size. For instance, clays may have a porosity of 50% with specific retention of 48%.

'Specific Yield' also referred to as 'Drainable Porosity' of an aquifer, is the specific quantity of water released by the saturated pores under the influence of gravity. It can be defined as the ratio of the volume of water, which after saturation, can be drained by gravity to its own volume. The volume of water retained and the volume of water drained constitute the total volume of the saturated porous material. Specific yield is a dimensionless parameter of an aquifer. The specific yield of a formation is a useful index, that determines the quantity of water that the formation would release under the influence of gravity, if tapped by a well.

'Specific Storage' of an aquifer is defined as the volume of water released from or taken into storage per unit volume of an aquifer, per unit change in hydraulic head.

'Intrinsic Permeability' is the ability of a porous medium to transmit a fluid through it. It is a property of the medium only and is independent of fluid properties.

'Specific Capacity' of a well is the index of productivity or the yielding capacity of a well. It is expressed by the ratio of discharge of the well in m^3/day and the drawdown caused in the pumping well in metres. In other words, it is the discharge (yield) per unit drawdown. It is expressed as, the pumping rate (Q) divided by the drawdown in metres, i.e. SC = Q/S where Q is the discharge (m^3/day) and S is the drawdown in metres. It can be determined as follows: if a well pumps at the rate of 1000 m^3/day and the drawdown is 10 m, then the specific capacity of the well is 100 m^3/day/m of drawdown. An increase in diameter increases the specific capacity as the yield of the well. Larger thickness of the aquifer provides a higher specific capacity. The specific capacity is independent of the duration of pumping. It is an useful parameter that can be helpful in classifying the wells in an area based on their yield. Higher specific capacity wells indicate higher transmissivity of the aquifer. The diameter of dug-wells in Goa varies from 2.2 to 6.1 m with a specific capacity varying from 10.6 to 228.7 m^3/day/m whereas the exploratory borewells (22.0–202 mbgl) have specific capacity of 0.27–988.5 m^3/day/m.

Permeability is the ability of a fluid to move through the pores of a medium. It determines how well connected the pores are within a rock or sediment. The permeability can be explained as the quantity of water (fluid) flowing through a unit cross-sectional area under unit hydraulic gradient. It is measured in m/day. It depends on the characteristics of the porous medium such as the hydraulic conductivity, density of water, the grain size and the viscosity of water at a given temperature.

Transmissivity of an aquifer is the quantity of water flowing through a vertical opening of unit area extending the entire thickness of the aquifer per unit hydraulic gradient.

4.4 Surface Water Resources

Surface water includes the meteoric waters (rainfall), rivers and impounded rivers. It is primarily contributed from precipitation which includes rainfall, snowfall, frost, hail, and the like. Of the total quantum of water, the global freshwater reserves constitute hardly about 3%, (Fig. 4.2) which amounts to 37.5 Mkm^3 of which about 23% is available for effective utilization, the balance is locked in glaciers and ice sheets.

India has barely 4% of the world's freshwater resources but hosts nearly 16% of the world's demography. With changing weather pattern and recurring droughts, the country is invariably water-stressed. Average rainfall in India is about 1215 mm; the west coast and north-eastern India receive about 3500–4000 mm of rain but it decreases to less than 600 mm and at places, to less than 100 mm in north-western India and interior parts of the Deccan plateau. In southern India, the sole contribution of water is through rainfall. Due to monsoon variability, average rainfall in the country is highly variable each year. The regions receiving high rainfall are affected by floods, while those of low rainfall are drought-prone.

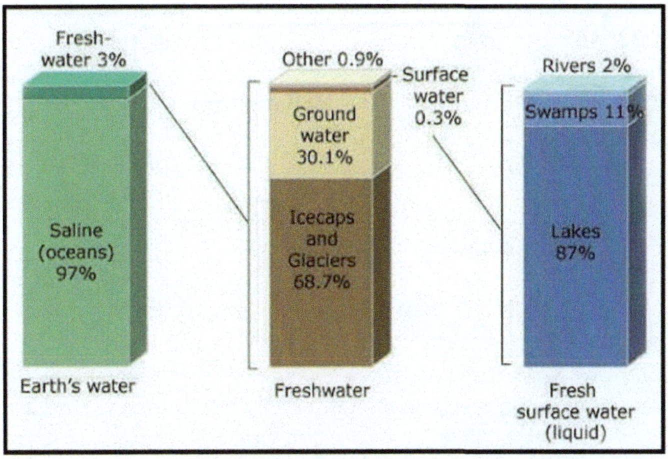

Fig. 4.2 Distribution of water on Earth (data source Shiklomanov [1993]; modified after United States Geological Survey [2019])

4.4.1 Surface Water Resources of Goa

The annual precipitation in the State varies between 3000 and 3800 mm. Of the total rainfall received, a part is intercepted by vegetation, some infiltrates into the soil, some is lost by evaporation, and a major part is lost as surface runoff. Presuming an average annual rainfall to be 3,500 mm (State area 3702 sq. km), the Central Water Commission (CGWB 2019) has calculated the water potential of the State to be 8570 Mm^3 of which nearly 40% as per CGWB (ibid.) and 80% as per NIO (Anon 2011) goes to the sea as runoff, after accounting for the storage for irrigation and urban water supply (ibid.). Irrigation accounts for 1465 Mm^3 of which 1125 Mm^3 comes from surface water and the balance 340 Mm^3 from groundwater. The quantum of runoff is approximately corroborated from the discharge by all the seven rivers, namely Tiracol, Chapora, Mandovi, Zuari, Sal, Talpona and Galgibaga together with two rivulets Baga and Saleri, taken together (Fig. 4.3).

More than 65% of the total discharge from the State is contributed by the Mandovi and Zuari rivers. A small fraction, ~152 Mm^3, of the total rainfall received contributes to the recharge to groundwater. The watershed-wise guesstimate of the contribution to the groundwater potential of the State and the dependable yield are presented in Table 4.1.

The combined water requirement of the state is 1329 Mm^3. The groundwater potential of the State is placed at 160.33 Mm^3 against the annual abstraction of 53.71 Mm^3 (Anon 2021). There is a lot of scope for conservation of water in the State (CGWB 2019).

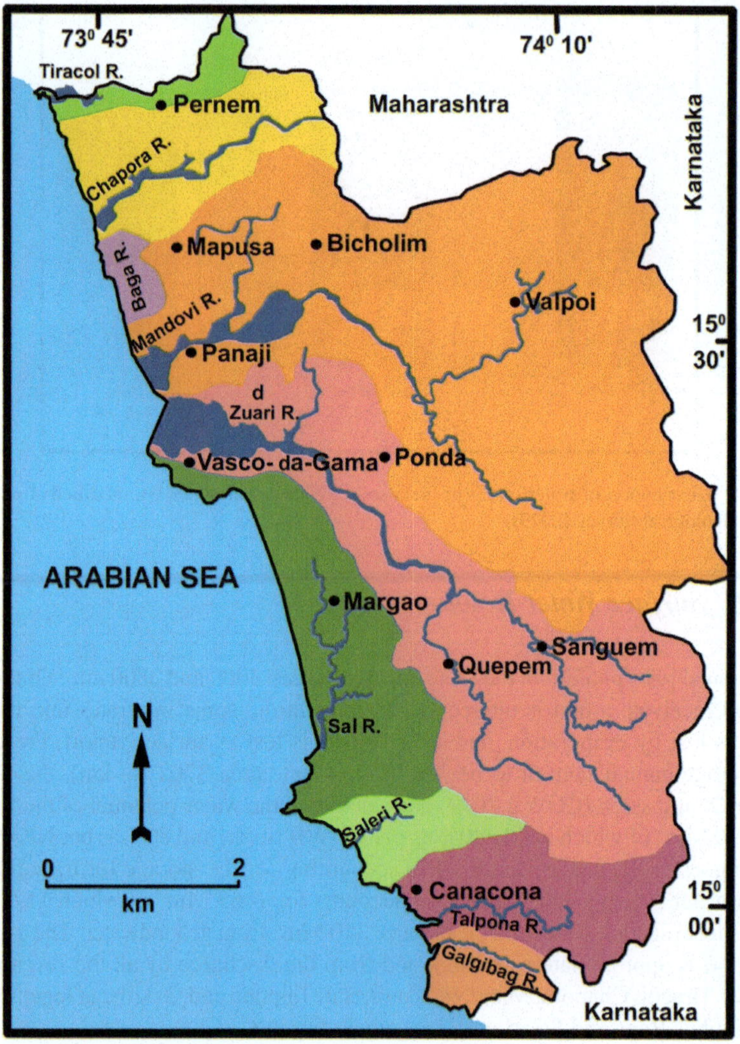

Fig. 4.3 Map depicting the watersheds of the major rivers of the State (after Water Resources Department, Government of Goa)

4.5 Groundwater Resources

The water beneath the surface in a broad sense can be referred to as groundwater although it is customary to refer to it as subsurface water. It is described as water located beneath the ground surface in soil pore spaces and, in the pores and fractures of lithological units beneath the soil zone. It is the largest source of fresh water for the humanity. As mentioned earlier, nearly 77% of fresh water available on the globe

Table 4.1 Surface runoff and the groundwater potential of various watersheds from the State (after Central Ground Water Commission, in CGWB [2019] and Anon [2021])

Sr. No.	Watershed	Catchment area (sq km)	Surface runoff (Mm^3)	Groundwater$^{\infty}$ recharge (Mm^3)	Dependable yield (Mm^3)
1	Tiracol	71	164.25	32.85	–
2	Chapora	255	588.35	117.67	–
3	Baga	50	116.42	23.28	–
4	Mandovi	1580	3580.04	716.00	1639
5	Zuari	973	2247.4	449.48	867
6	Sal	301	694.49	138.89	–
7	Saleri	149	343.04	68.60	–
8	Talpona	233	515.59	103.11	231
9	Galgibaga	90	187.11	37.42	86
	Total	3702	121.386	1687.30	28.23

Note $^{\infty}$Computed guesstimate 20% of the surface runoff

is locked up in polar ice caps, of the remainder 30% constitutes groundwater whereas the rivers and lakes merely represent 1%.

The demand for water has been growing in recent years largely as a result of demographic growth, industrial and agricultural boost, deterioration in quality of surface waters and reduction in availability of surface sources in case of prolonged droughts. Ever-increasing demand for water has necessitated a more systematic exploration and sustainable development of the groundwater resources. Nearly half the national urban water requirement and nearly similar quantum for irrigation is met through groundwater. It contributes to about 80% of the drinking water requirements in rural areas.

4.5.1 State Groundwater Resources

The groundwater regimen of the State, for the sake of simplicity and convenience, is described with reference to three hydrological units which correspond broadly to the three lithological types. These are: (i) the beach sands and alluvial-colluvial deposits viz. the fluvial sediments and intermontane valley deposits, (ii) the laterites which essentially include the lithomarges (saprolites), dominantly clayey-silty in nature, that occur beneath the duricrust-tablelands (palaeoplain surfaces) and (iii) the metasediments and granitic gneisses. Field observations, exposed sections of rocks in mine pits and dug-well sections, permit categorization of the groundwater aquifers in the State into three lithological types: (i) the lithomarges which form the unconfined aquifers, (ii) the sandy alluvial-colluvial deposits of coastal plains that also occur as unconfined aquifers and laterally abut against and often grade into the lithomarge aquifers and (iii) the weathered country-rock metasediments/granitic gneisses which often occur as the confined aquifers.

A typical section of the tableland consists of 5–25 m thick laterite duricrust at the top which is followed by a thick zone of lithomarge (refer Fig. 5.1). The latter varies in thickness from 30 to >70 m in which case it extends even below the MSL. This horizon is followed by the partly weathered and fractured parent rock represented by the metasediments or granitic gneisses. The duricrust does not have sufficient water-holding capacity due to both, the inherent and superimposed permeability from fracturing, as also its geomorphic location. The lithomarge horizon despite being largely made of clay minerals (kaolinite, halloysite, degraded chlorite) is capable of storing and transmitting water. At places, the lithomarges contain bands and lenses of metasediments, such as the sericite quartzites, banded haematite quartzites, magnetite quartzites and the like. The partly weathered and fractured parent rock occurs beneath the lithomarge.

Groundwater occurs under two situations: (i) unconfined conditions, and (ii) semi-confined conditions. Under the watertable conditions, the main groundwater aquifer occurs in the weathered zone below the lithomarge horizon. This is also the case of beach sands and alluvial sands. The presence of intercalated clays at places gives rise to sub-artesian conditions.

The occurrence and movement of groundwater in the metasediments are along the permeable zones developed due to weathering and fracturing of rocks. In the upper zones, the movement of water is largely through the permeable weathered rock, whereas in the lower, relatively less weathered portion, the movement is along fracture zones, joints and fissures. The transition zone between the lithomarges and the weathered basement rocks forms a confined aquifer. Often this aquifer occurs below the MSL.

Semi-confined conditions of groundwater are encountered in the weathered zones below the laterite duricrust and in gravel zones beneath clay partings within beach sands. The movement of groundwater takes place essentially through joints and fractures of the duricrust until the depth of the lithomarge horizons. The lithomarge horizons in some cases serve as local confining layers that prevent the downward movement of groundwater. These aquifers at places show continuity with those from the coastal sandy plains, as often the sands occur in juxtaposition with the weathered zones of the laterite profiles.

The coastal plains including the stabilized dunes and the alluvial-colluvial deposits are characterized by primary porosity. The movement of groundwater is due to the inherent permeability of the rocks. The groundwater occurs in the intergranular sand and silt that are intercalated with beds and lenses of gravel and pebbles. Clays, muds and silty clays occur as intercalations with the sands. These serve as limited aquicludes.

The hydrological data for the three lithological units are presented in Table 4.2 (after Adyalkar 1985).

The aquifer parameters of some selected wells in different hydrological settings have been investigated (Adyalkar 1985). The details of the pump tests conducted in 12 open wells are presented in Table 4.3 (ibid.).

Table 4.2 Depth of water level in wells from the different lithological units

Sr. No.	Hydrological unit	% of wells inventoried	Depth range (m)	Depth to the water level (m)
1	Beach sands and alluvial-colluvial deposits	16.6	1.8–8.5	0.5–7.2
2	Laterites and weathered zones	60.7	2.1–20.1	0.5–18.8
3	Metasediments and granitic gneisses	22.7	2.8–14.4	1.4–13.1

Table 4.3 Aquifer parameters of selected wells from the different lithological units (after Adyalkar 1985)

Sr. No	Aquifer	Saturated thickness (m)	Specific capacity lpm/m of dd	Hydraulic conductivity m/day	Specific yield (%)	Radius of influence	Transmissivity sq m/day
1	Alluvial sand	1.20	101.50	99.20	–	–	119.40
2	Beach sand	2.2–2.40	46.20	10.00	1.10	17.52	22–24
3	Laterite (underlain by weathered gneiss/schist	2.0–2.50	48.20	13.80	4.10	–	22.60–34.50
4	Laterite/ lithomarge	1.1–2.80	262.00	109.40	3.5–4.10	11.62	120–306.32
5	Weathered granite	1.40	169.00	57.60	–	–	80.64

4.5.2 District-wise Groundwater Distribution

An estimate of the groundwater availability is presented in Table 4.4 (after CGWB 2019).

The district-/*taluka*-wise distribution is suited for administrative purposes when dealing with the management of the water resource requirements of the State. The National Water Policy, revised in 2012, recommends that the watershed should be the unit for all water resources planning. When dealing with academics of water such as the science of water, groundwater movement, distribution and groundwater development, it is more appropriate to discuss the variations with respect to a watershed, since the latter is the basic unit in hydrology that enables us to understand the close interrelationship of water, the surface morphology of the basin, its relationship with the subsurface geology (Fig. 7.2) and the structural fabric of the rocks, all of which influence the storage potential of the aquifer. In more specific terms, the watershed allows us to discuss the distribution and movement of groundwater within the different watersheds as it takes into account the surface morphology and the hydrology of the watersheds. These factors have a considerable influence on the storage potential of

Table 4.4 District-wise groundwater resource availability and demand of the State (after CGWB 2019)

	District	North Goa (Mm^3)	South Goa (Mm^3)	State Total (Mm^3)
Monsoon	Recharge Rainfall	84.87	49.80	13,467
	Recharge Other sources	0.51	7.86	8.37
Non-monsoon	Recharge Rainfall	0	6.44	6.44
	Recharge Other sources	44.64	27.10	71.74
	Total	130.03	91.21	221.23
	Natural discharge (non-monsoon)	52.01	36.48	88.49
	Net GrW[@] availability	78.02	54.73	132.74
	Irrigation	8.42	5.13	13.55
	Domestic/industrial	17.04	13.23	47.31
	Total (district)	25.47	18.37	43.83
	Projected demand (domestic/industrial)	21.10	16.39	37.49
	GrW[@] availability for future (irrigation)	48.49	33.20	81.69

Note [@]GrW: groundwater

the terrain. However, in the absence of watershedwise groundwater data of the State, the district/*taluka* level data are utilized in this narrative.

The *taluka-wise* groundwater potential and development are presented in Table 4.5.

4.5.3 *Spatiotemporal Variation of Groundwater*

With growing demography, rapid urbanization and development, and increasing demand for water have impacted the watertable conditions in some areas of the State. The watertable being a dynamic entity shows fluctuation in depth depending on the pace of recharge and the rate of abstraction from the aquifer. The depth of fluctuation depends on several factors such as the local hydrogeological conditions, frequency, duration and quantum of rainfall, topography of the region and extent of groundwater development. The CGWB monitors both North and South Goa districts four times a year with over 25 hydrograph stations in each district (CGWB 2013). The seasonal (pre-monsoon) depth variation of the two main aquifers in the year 2016 ranged from more than one to nearly 19 m below ground level (bgl) with maximum water level recession (5–10 m) in Bicholim and Ponda *talukas* (Figs. 4.4 and 4.5). In

Table 4.5 *Taluka-wise* groundwater development scenario in the State from 2011 to 2017 (after CWC in CGWB 2019)

Taluka	GrW potential (Mm^3)	GrW draft (Mm^3)	Dev. stage (%)	GrW potential (Mm^3)	GrW draft Mm^3	Dev. stage (%)	GrW Potential (Mm^3)	GrW draft (Mm^3)	Dev. stage (%)
	2011			2013			2017		
Pernem	16.17	4.65	35	14.90	6.12	41	13.71	6.34	46
Bardez	17.73	7.68	43	16.54	8.15	49	23.44	7.06	30
Bicholim	17.62	3.90	22	19.13	3.50	18	24.64	3.43	14
Sattari	8.16	2.65	32	73.2	4.25	58	15.10	3.15	21
Tiswadi	13.25	4.65	35	14.90	6.12	41	13.71	6.34	46
Mormugão	12.840	3.60	28	12.86	3.84	65	12.95	4.07	31
Ponda	12.59	3.82	30	10.43	4.77	45	11.56	5.35	46
Dharbandora	–	–	–	5.21	1.04	20	5.88	1.26	21
Salcete	16.79	7.55	45	15.04	9.79	19	25.14	13.12	51
Sanguem	14.28	1.65	12	7.31	1.46	15	8.07	2.52	31
Quepem	10.09	1.77	18	12.04	2.23	23	7.87	3.07	39
Canacona	5.87	1.26	22	5.48	1.25	20	3.48	1.09	31
Total	**145.43**	**4114**	**28**	**146.25**	**53.77**	**37**	**160.33**	**53.71**	**34**

Note GrW: groundwater, Dev: development

the post-monsoon season, the decline varied from less than a metre to a maximum of over 14 m, especially in the southernmost part of the State (Figs. 4.6 and 4.7).

When these values are compared with those during the previous decade (2005–2015), majority of wells in the northern, western and central parts of the State showed recession in watertable in the pre-monsoon season (Fig. 4.8). The northern district, in particular, showed watertable decline even in the post-monsoon season (Fig. 4.9, Table 4.5).

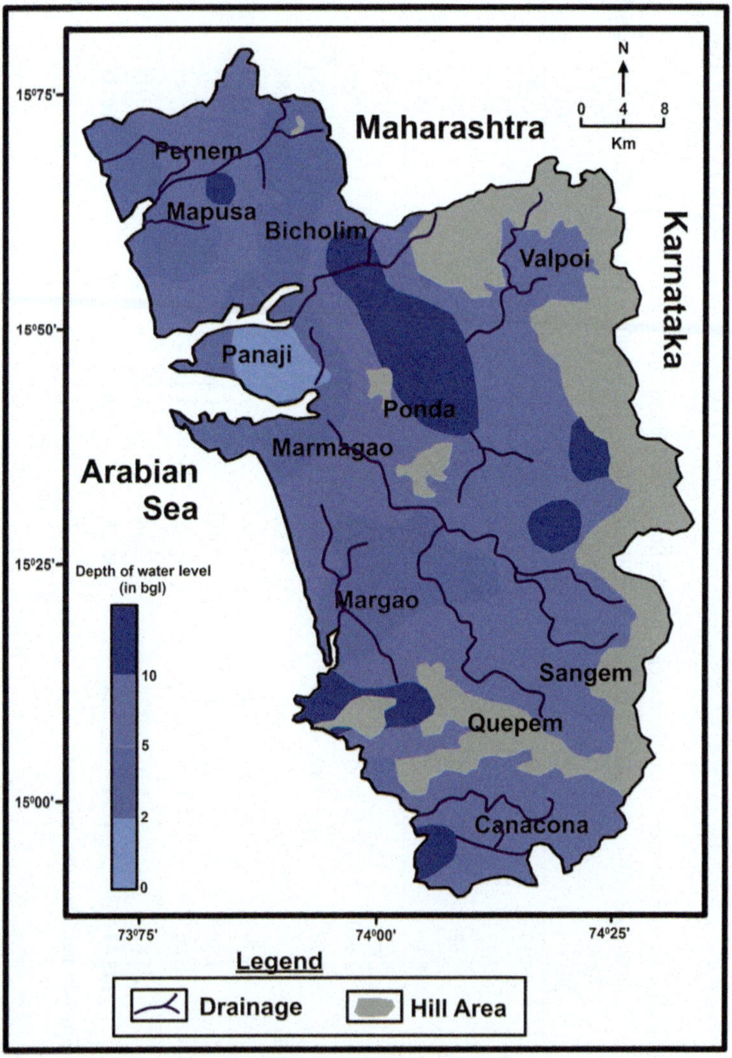

Fig. 4.4 Pre-monsoon (May 2016) depth to water level map of aquifer I (modified after CGWB 2019)

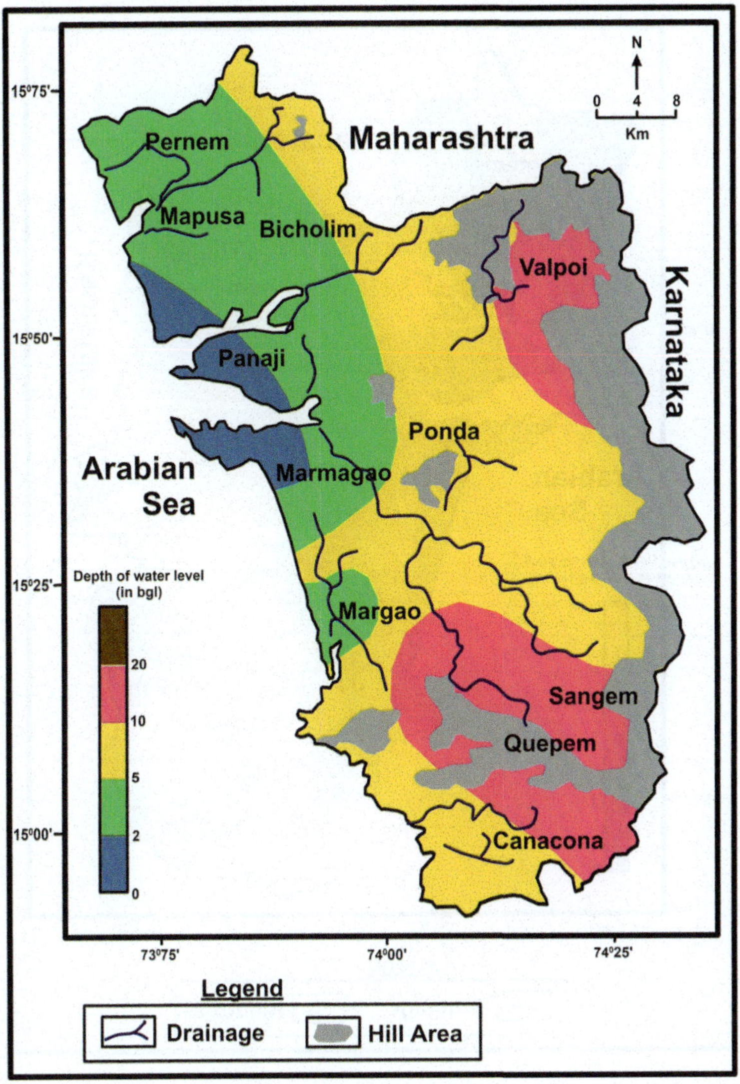

Fig. 4.5 Pre-monsoon (May 2016) depth to water level map of aquifer II (modified after CGWB 2019)

4.5.4 Sustainability vs Resource Depletion

Although sustainability assessment involves several factors including potential aquifer storage, recharge capacity, the quantum of withdrawal among others, based essentially on the annual groundwater recharge and annual groundwater draft (consumption/withdrawal) a sustainability assessment of groundwater has been made

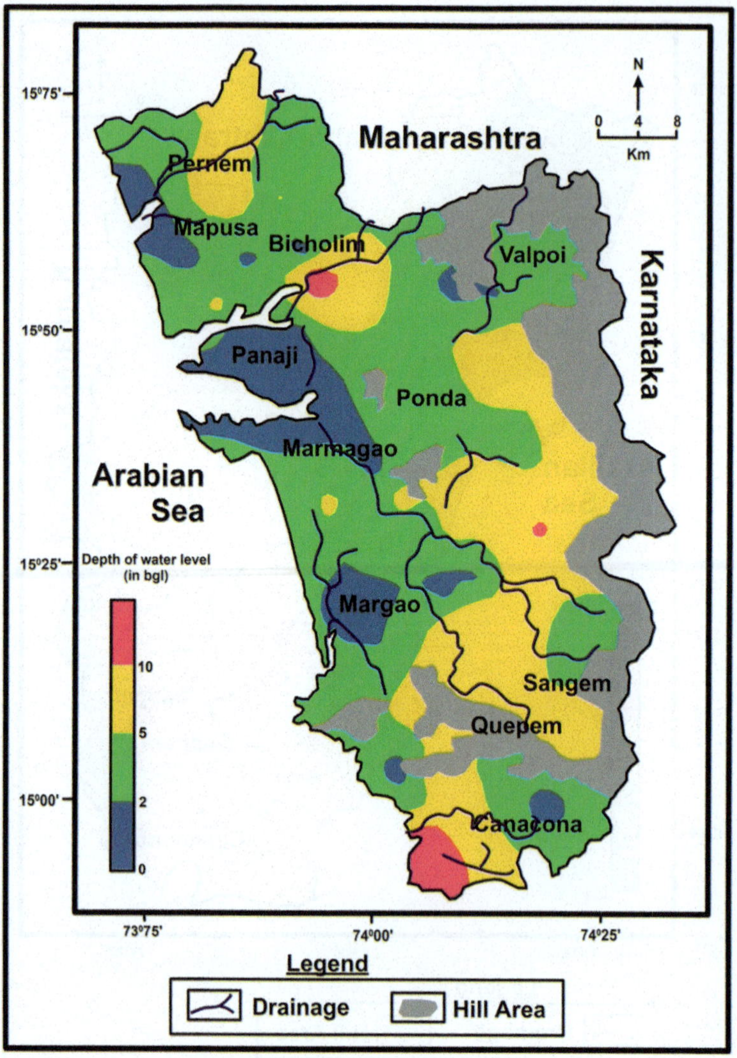

Fig. 4.6 Post-monsoon (November 2016) depth to water level map of aquifer I (modified after CGWB 2019)

(CGWB 2019) which could be improved upon with further work. The declining trend of the groundwater table is a cause of concern which calls for better monitoring and regulation to prevent further decline. There could be two main reasons for the fluctuations in the watertable: (i) excessive withdrawal of groundwater and (ii) inadequate recharge to the aquifer. The declining trend of water levels especially in the post-monsoon season suggests that the aquifer is being desaturated due to

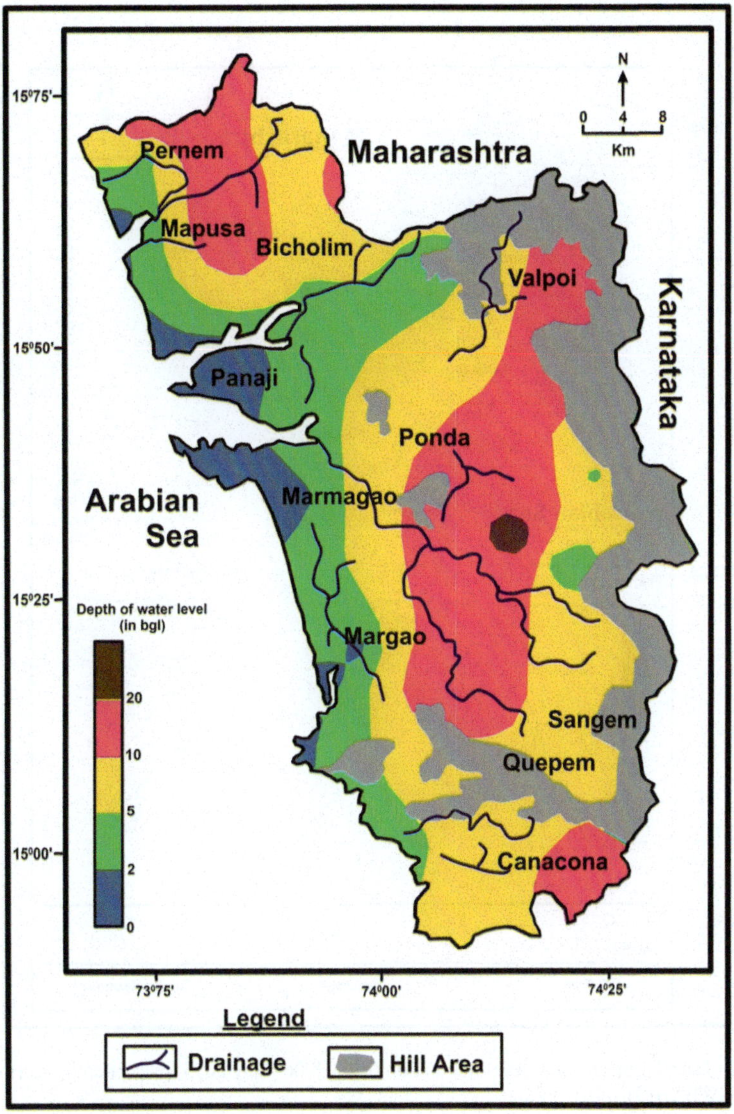

Fig. 4.7 Post-monsoon (November 2016) depth to water level map of aquifer II (modified after CGWB 2019)

large-scale abstractions or is not getting adequately recharged and hence is progressively getting depleted. It is difficult to pinpoint precisely the reasons for the decline unless a detailed study of the aquifer characteristics and well inventory of a majority of wells is undertaken at an appropriate scale. Yet, it is possible to broadly discuss the potential reasons for this anomalous situation. One of the reasons for the decline could be related to the topography of the region (refer Sect. 1.3). The elevated eastern

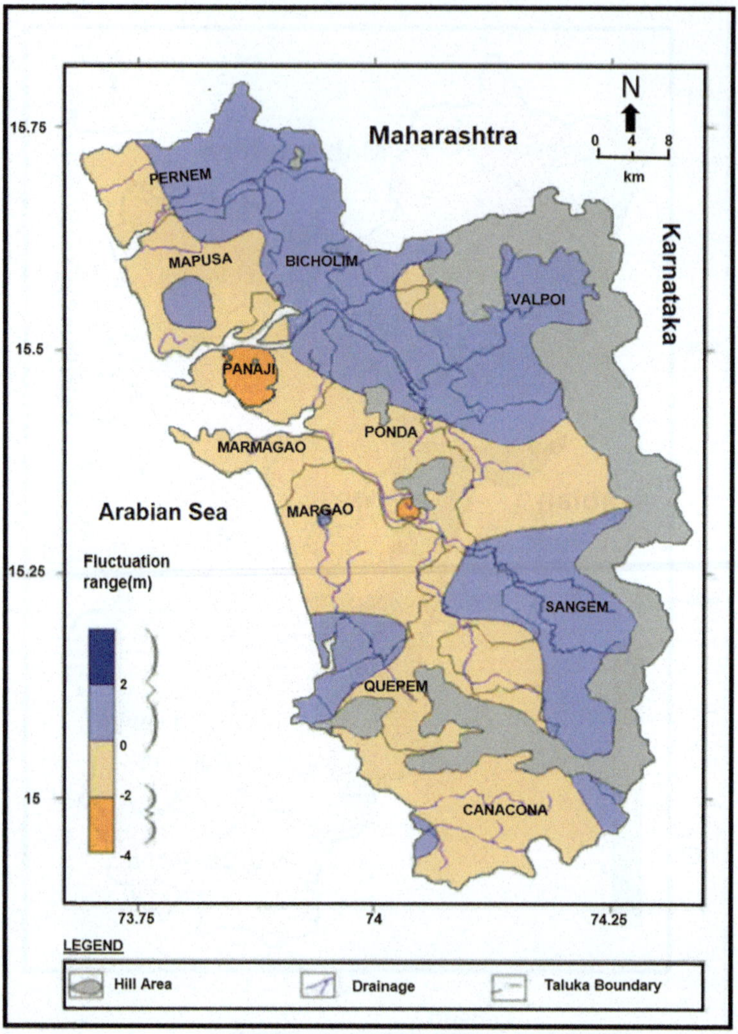

Fig. 4.8 Decadal mean water level fluctuation (May 2006–May 2015) map of aquifer I (modified after CGWB 2019)

region of the State contributes higher recharge to the low-lying western part. The evidence in support of this observation is provided by the higher altitude wells going dry earlier in the post-monsoon season than in the previous decade, despite the longer duration of post-monsoon rains in the current decade.

The second possible reason could be the longer post-monsoon season rains experienced in the current decade than in the previous one. The longer season contributes to larger recharge to the higher altitude areas that revives the low altitude aquifers in the west from post-monsoon contribution.

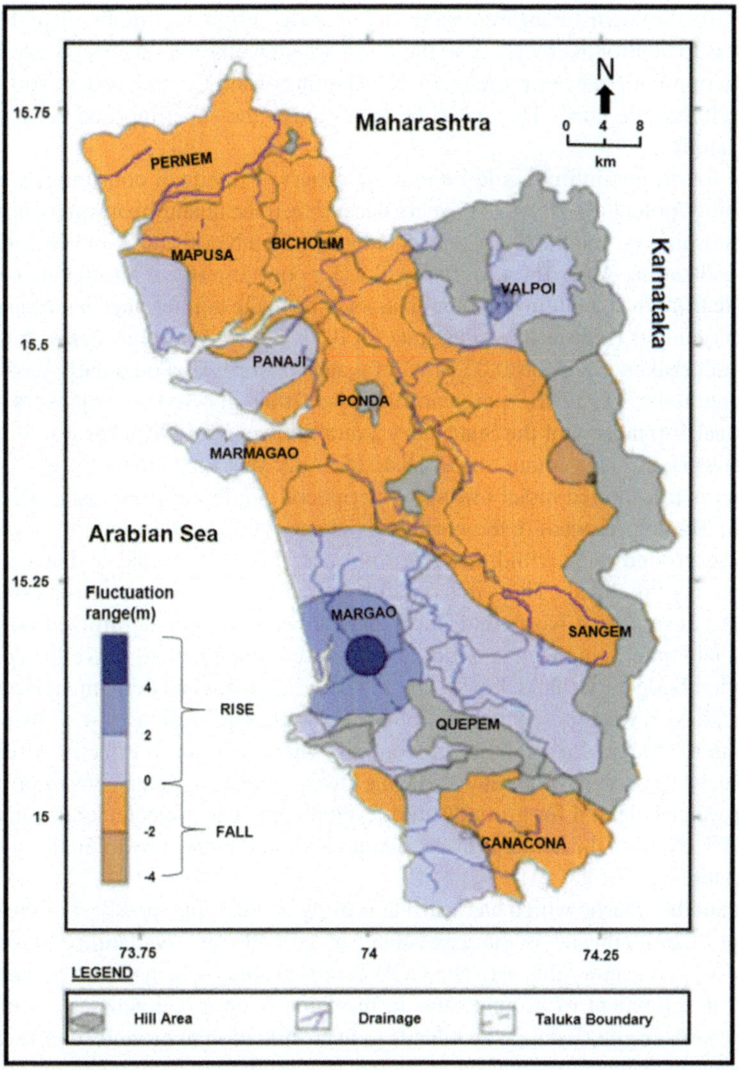

Fig. 4.9 Decadal mean water level fluctuation (Nov. 2006–Nov. 2015) map of aquifer I (modified after CGWB 2019)

The third reason could be the larger contribution of secondary recharge to the aquifer due to the construction of *bunds* (embankments), *bandharas* (weirs)/vented dams, check dams, dams, percolation tanks and other similar structures in recent times. Although these structures are normally constructed to augment storage and are not intended to increase percolation per se, they have been responsible to effectively enhance the secondary recharge. As, for instance, in the case of Salaulim dam in south Goa, it has been established using tracer techniques that secondary recharge in the downstream areas increased considerably post-completion of the dam. Attempts to control the recharge to the downstream wells and springs failed, despite the grouting undertaken on the recommendations of the Central Waters and Power Research Station, Pune. This could be more due to the superimposed permeability of the aquifers.

The fourth possibility could be that the observed trends of consumption versus availability/potential of the resource are deceptive, if the quantum of water consumed by the hospitality, and the construction industry of the North and South Goa districts is any indication. An estimate of the total water consumption of North Goa needs to be made from the quantum of water abstracted from the aquifer plus that transported by water tankers (registered in the State for the water-transport business). A total of 605 water tankers transport 0.03 Mm^3 of water to North Goa on a daily basis. This works out to be 10.95 Mm^3 per year. Likewise, 700 registered water tankers and an additional 100 tankers of the State carry a minimum of 0.045 Mm^3 of water daily to various working sites, amounting to 16.42 Mm^3 per year to South Goa. This quantity of water is transported mostly from the peripheral *talukas* to the coastal belt of the Pernem, Bardez, Tiswadi in the north and Salcete *taluka* in the south.

These groundwater withdrawals are not accounted for in the budgetary water projections of the State. Thus, it is possible that as a result of inputs from elsewhere the water resource of North and South Goa appears to be underutilized on paper, although in practice a much larger quantum is consumed, though not reflected in the utilization/budgetary charts. For instance, Tiswadi, Bardez and Pernem *talukas* taken together have a water potential of 45.68 Mm^3 whereas utilization is 14.17 Mm^3. The quantum of 10.95 Mm^3 brought in from elsewhere and put to effective use is not adequately represented, thereby showing lower figures of water consumption and higher water potential for groundwater development. The projected consumption is only 31% whereas the actual is 56%. A similar situation may prevail in the southern district too.

Yet another reason which has a strong bearing on the future prospect of the development of groundwater is the assessment of groundwater availability based on a limited well-inventory dataset. The CGWB (2019) study which shows potential for further development of groundwater in the State is on a regional scale with very limited well-inventory data which leads to overestimation of groundwater potential. If a larger well-inventory dataset collected at a large scale were taken into consideration in calculating the quantum of groundwater availability, a different picture of groundwater development may emerge.

Inferences on the Future Scenario: The total annual recharge to the groundwater is estimated to be 284.71 Mm^3 and the annual groundwater availability is placed at 267.12 Mm^3, the development is about 27% in comparison with states, such as Haryana where the development is 112%. The groundwater draft for irrigation and industrial supply is 32.66 Mm^3. The utilization of groundwater is barely 8.2%. However, decadal fluctuations in the levels of groundwater table are indicative of the regional depletion of the resource. For instance during the period from May 2006 to May 2015, the water levels recorded a recession from −2 to −4 m (CGWB 2019) in more than 60 % of the wells from aquifer I. The decline in the groundwater level seen over time is a matter of concern as it can lead to the water scarcity, deterioration of groundwater quality and the overall degradation of the environment. Such a scenario may necessitate expensive and habitat destroying water projects in future.

Moreover, the inferences drawn on the quantum of water available from a limited dataset (CGWB 2019) need to be taken with caution, since reliable and requisite amount of the details of well-inventory data on the desired scale are not available for the State as a whole. Furthermore, excessive withdrawals of groundwater are localized to certain specific areas, for instance, the western coastal belt which broadly includes the 4 metropolitan towns, Vasco da Gama, Mapusa, Panaji and Margao where nearly half the population lives, whereas the rest of the State is not subjected to such withdrawal pressures. It is advisable that the data with excessive localized withdrawals are avoided for the purpose of generalizations. The present groundwater potential of the State is 160.33 Mm^3, the groundwater draft is 53.71 Mm^3 whereas the combined water requirement is placed at 1329 Mm^3 (ibid.). Considering these factors, it appears a little premature to conclude that there is scope for further groundwater development in the State by digging new wells.

In the light of the existing data, it would be more prudent to infer that further groundwater development in the State should be attempted with due caution (also refer Sect. 4.6.2). More emphasis should be placed on water conservation and reuse of used water, rather than merely developing the existing resources and searching for new ones. A detailed and systematic groundwater assessment of the State needs to be undertaken by systematic delineation and mapping of aquifers on a much larger scale followed by systematic aquifer characterization before any concrete recommendations are made as regards further development of groundwater. In the meantime, caution needs to be exercised in the utilization of groundwater by adopting adequate conservation measures and following sufficient aquifer recharge procedures with adequate safeguards. Finally, it may be inferred that the groundwater situation in the State is far from sustainable levels, and certain mitigational measures need to be adopted on a priority basis.

4.5.5 *Mine Dewatering: Implications on the Watertable*

The stress on groundwater aquifers depends on the groundwater development within the watershed. It is primarily governed by the difference between the quantum of water extracted and the amount of total recharge. Accordingly, watersheds can be classified into four categories: (i) non-critical (less than 50% of utilizable water is withdrawn), (ii) sub-critical, 50–75% of the water is extracted, (iii) critical 75–100% of the utilizable water is extracted and (iv) most critical, when the extraction exceeds the total utilizable water (CGWB 1984). Sustainability, therefore, largely depends on the extent of the extraction rate to the replenishment rate, save issues such as prevention of saltwater intrusion, pollution, land subsidence, soil salination among others.

Normally, a dynamic balance is maintained between recharge and discharge where undisturbed conditions exist. Broadly withdrawal of water for normal sustenance including agriculture and domestic consumption may not affect the groundwater balance appreciably, since withdrawals are compensated by the recharge from rainfall. However, anthropogenic activity such as excessive mining can lead to disturbance of the groundwater balance. This could be the reason for the decline of the watertable observed in the mining belt of north and central Goa where both quantity and quality of groundwater have been variously impacted depending on the intensity of interference with the groundwater regime.

Recession of Watertable: The Bicholim, Sattari and Sanguem *talukas* host the mining belt of north and central Goa. These *talukas* exhibit a recession of watertable expressed as an NW–SE-directed anomaly (Figs. 4.4 and 4.9; CGWB 2019) which possibly suggests a direct connection with overexploitation of groundwater. A number of iron ore mines aligned almost in the NW–SE direction occur in this area. The majority of mines are deep and dewatering of the mine pit is essential to exploit the ore body beneath the local watertable. Although the figures for the total quantum of water pumped out from the entire mining belt are not available, a guesstimate for the entire belt could be made from the quantity of water pumped out from one of the major mines (Bicholim mines) from the area.

The quantum of water pumped out from a single pit of Bicholim mine for instance ranged from 120 to 180 m^3/h and continuous pumping is resorted to intervals varying from 227 to 546 h/month (Chaulya et al. 2000). Several mines are involved in the dewatering process. The quantum of water pumped out is enormous. Field observations around the mines show that wells in topographically higher elevation show drying of dug-wells (~30% of wells) and springs tapping this unconfined aquifer from the mining belt. The recharge to groundwater is also restricted in these higher altitude areas.

However, according to a recent study conducted in the area (ibid.), dewatering of the mines does not affect the watertable and the dug-wells in the surrounding region/village as there is no noticeable connection among the open mine pits and the dug-wells in the nearby villages, as both are governed by independent sub-watersheds, formed due to the presence of impervious lithomarge (clay) partings within the aquifer in the lateritized zone. Moreover, the dug-wells penetration is shallow (merely the laterite duricrust, depth 4–6 m) and tap the unconfined aquifer at the transition of the duricrust-laterite and the lithomarge below, whereas the mining operations are much deeper (depth >30 m) below the duricrust/lithomarge transition zone and into the weathered phyllites/BHQ, i.e. the deeper aquifer. Such situations, wherein there is no effect on dug-wells from mine dewatering, may be true in case of some selected mines, however, this does not appear to be the case for the entire mining belt.

Water Scarcity due to Mine Dewatering: To the SE of Bicholim for instance, there are a number of villages, e.g. Aravalem, Honda, Pissurlem, Kudnem, Bhuipal and others in the catchment of Kudnem River (tributary of Valvon, Mandovi watershed) within the mining belt which have experienced severe water scarcity due to dewatering of mines, to the extent that drinking water is supplied by water tankers.

The total amount of water pumped out merely from three mines is estimated to be 6.10 Mm^3. This quantum of water accounts for 44% of the groundwater storage (Choudri and Chachadi 2006). The withdrawal of such a large quantity of water has affected the base flow of the area leaving dry the river flow, ponds, springs and even soil due to lack of or diversion of base flow.

Impact of Groundwater Development on the Watertable: The recession of watertable as a consequence of groundwater development is noticed elsewhere in the State also. The borewells sunk to levels deeper than the dug-wells, to tap the lower aquifer have an adverse effect on the dug-wells. The dug-wells usually tap the phreatic shallow aquifer that occurs below the laterite duricrust. Excessive pumping of water from the bore well located in the vicinity of the dug-well has resulted in the drying of the dug-well due to induced contribution of water from the shallow aquifer to the deeper (lower) aquifer. Such instances have been noticed at Assagao Bade where pumping at the rate of 200 m^3/day has resulted in drying of the dug-well located about 70 m from the borewell. Likewise, at St. Michel Wado (Dando) at Anjuna, a dug-well had to be deepened, after a bore well was sunk in its vicinity. The impact of excessive pumping from the borewell on the dug-well needs to be ascertained by conducting more pumping tests not only here but also in other parts of the State. As a precautionary measure therefore, new wells are not allowed to be dug within 100 m from the existing wells (*Times of India* dtd. 24.02. 2020).

Positive Impact of Mine Dewatering: Large-scale dewatering of mine pits has also had a positive impact on the groundwater balance due to changed recharge regimes and water-use practices. A quantitative estimation of the groundwater recharge of the mining belt has been carried out on the bases of the watershed area, groundwater level fluctuation and the specific yield of the aquifer (Choudri and Chachadi 2006) despite limitations of this approach. Over 70% of the wells from the mining belt, however, exhibited varying increase in groundwater levels. Such a trend is commonly seen in the mining areas where there is excessive dewatering of the mine pits to win the ore beneath the watertable. The likely reasons for the increased levels of groundwater could be: (i) the water pumped out for dewatering of mine pits, although partly utilized, is ultimately let out to the natural drains such as the stream courses. It contributes to additional recharge to the unconfined aquifer by return flow. (ii) The abandoned mines and exhausted mine pits store large quantities of water, at times utilized for pisciculture, recreational purposes and also used for water harvesting. This water contributes to secondary recharge. (iii) The mine rejects which normally exceed three times the ore mined, retain sufficient quantity of rainwater which also provides secondary recharge to the groundwater. In addition, settling ponds and silt-traps as also exploratory drill-holes contribute to additional recharge to groundwater regime (Table 4.6). A small percentage of wells, however, do not show any change in groundwater levels as the total recharge almost compensates for the discharge and the continuity of the watertable is not maintained due to the presence of impervious lithomarge.

Groundwater Availability versus Utilization: The groundwater utilization figures of South Goa when seen in the context of the water potential of the aquifer suggest

Table 4.6 Water level trends in the State during the period 2001–2010 (modified after CGWB 2013)

Sr. No.	*Taluka*	Pre-monsoon		Post-monsoon	
		Fall (m)	Rise (m)	Fall (m)	Rise (m)
1	Pernem	0.2–1.5	0.4–1.5	0.04–1.4	0.0–0.67
2	Bardez	0.0–0.0	0.2–3.5	0.0–0.12	0.1–1.8
3	Bicholim	0.3–7.5	0.0–0.4	0.0–0.16	0.6–3.5
4	Sattari	0.2–1.3	0.2–1.1	0.19–0.22	0.06–1.2
5	Tiswadi	0.0–0.0	0.2–1.9	0.0–0.06	0.03–2.2
6	Ponda	0.0–0.2	0.0–0.2	0.0–0.2	0.0–0.0
7	Salcete	0.5–0.7	1.5–2.0	0.0–0.8	0.0–0.1
8	Sanguem	0.1–0.7	0.1–0.4	0.2–0.4	0.0–0.4
9	Quepem	0.1–2.8	0.0–1.4	0.0–0.2	0.1–8.2
10	Canacona	0.0–0.02	0.0–1.6	0.1–0.8	0.2–0.4

sufficient availability of groundwater for future irrigation, since agriculture is uneconomical and hence is not practised in the State to the scale to which it used to be during the previous decades. This is evident from the large tracts of agricultural lands lying fallow. However, the observation of sufficient availability of groundwater appears constrained due to the non-availability of sufficient and reliable well-inventory data at the appropriate scale, as mentioned earlier, both on the aquifer characteristics and on the real-time abstractions.

4.6 Issues Related to the Salinity of Water

Background: Saline water commonly referred to as salt water is the water that contains high concentration of dissolved salts and other minerals often expressed in parts per million (ppm) of equivalent sodium chloride (NaCl). The US Geological Survey classifies saline water in three salinity categories: (i) slightly saline containing 1000–3000 ppm (0.1–0.3%), (ii) moderately saline 3000–10,000 ppm (0.3–1.0%) and highly saline water 10,000–35,000 ppm (1.0–3.5%), equivalent to 35 g of salt per one litre of water. Seawater has a salinity of about 35,000 ppm. Generally, water containing up to 0.05% of salt (0.5 g per litre of equivalent NaCl) is included under fresh water. That containing from 0.05 to 3.0% [0.5–30 g of salt per litre, often expressed as 0.5–30 parts per thousand (‰) with specific gravity between 1.0 and 1.02] is referred to as brackish water and that containing from 3.0 to 5.0% as saline. Brackish water may result from mixing of seawater with fresh water as in estuaries, or it may occur in brackish fossil aquifers as a result of connate seawater trapped within alluvial materials. Thus, the term brackish water covers a range of salinity and hence is not considered a precise term to express salinity conditions. In general, water with salt concentration of greater than 3.0% is considered saline, that which contains more than 5% and up to 26–28% of salts is categorized as 'brine'. There exist water bodies such as natural lakes which have higher concentration of salt than seawater, such lakes are referred to as Hypersaline lakes. Similarly, there are pools of water in the oceanic regimes with salinity levels three to eight times greater than the surrounding ocean waters, such pools are known as Brine pools. Salinity of water can be measured by determining its thermal conductivity. The thermal conductivity of seawater (3.5% of dissolved salt by weight) is 0.6 W/mK at 25 °C. It decreases with increasing salinity and increases with increasing temperature.

The western seaboard of the State has fairly extensive estuarine stretches where saline waters extend into the riverine stretches (Fig. 4.10).

4.6.1 Saltwater Intrusion

The coastal belt of the State is one of the sensitive areas as far as groundwater development is concerned. One, there is natural drainage of groundwater to the sea. Two, the coastal zone is the most densely populated region of the State with a state-average population density of 394 per sq km against a national average of 80 per sq km which is twice the world's average population density. Due to higher demographic density, improvement in living standards, enhanced construction, much denser industrialization and at places, improved agriculture, there is a distinct increase in demand for fresh water in the coastal areas. Groundwater, being the primary source of fresh water, is exploited at times even indiscriminately to meet the increased water demand (e.g. Hamed et al. 2018). As too much water is pumped out of the aquifer system, saltwater migrates landwards; this process is referred to as 'saltwater intrusion'.

If a pumping well is located close to the freshwater-saltwater interface, salt water may enter the well and contaminate the fresh water. Thus excessive withdrawal of fresh water leads to disturbance in the hydrodynamic equilibrium between fresh water and seawater existing in the aquifer. This invariably results in the upward movement of seawater (van Camp et al. 2014) which often underlies fresh water in these transitional aquifer areas. The ascendance of seawater causes depletion in the availability of fresh groundwater in the coastal aquifers.

4.6.2 Groundwater-Related Salinity Issues

The vertical movement of seawater in coastal aquifers has been well studied in recent years. It has been established that for a one-metre increase in watertable the thickness of seawater reduces by about 40 m. However, a reduction in the groundwater level below the MSL causes reversal of hydraulic gradient leading to the movement of salt

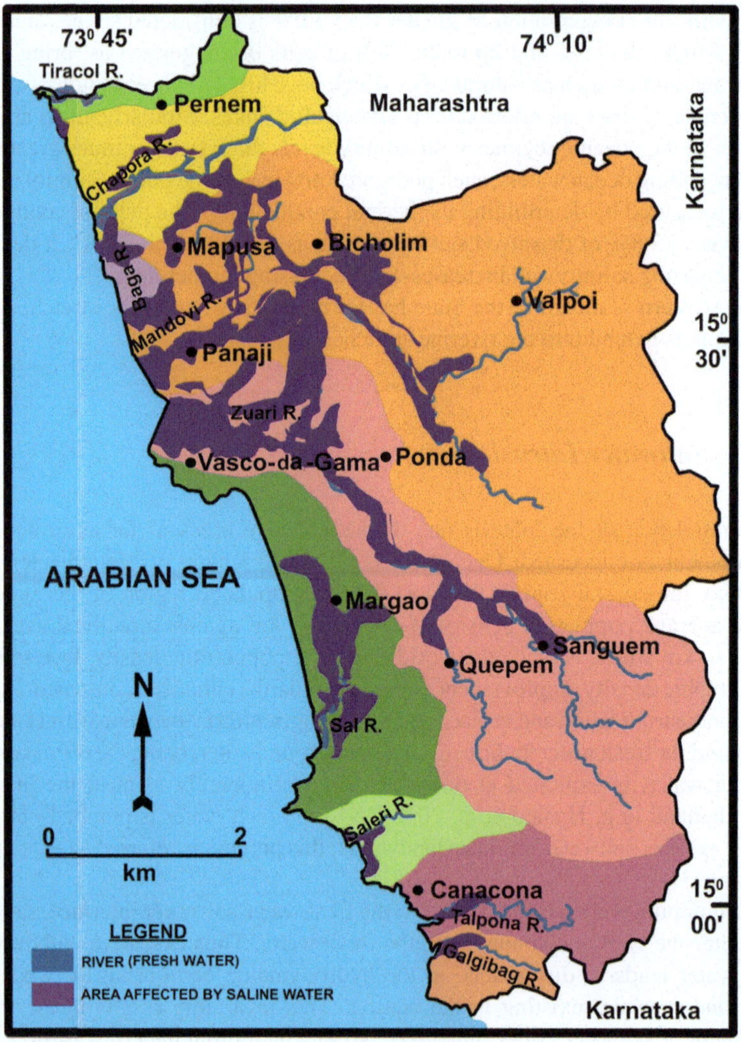

Fig. 4.10 Map depicting the estuarine stretches and adjoining salinity-affected areas of the various watersheds from the State (modified after CGWB 2013)

water into the coastal aquifers. This is generally referred to as 'Seawater Intrusion' (Fig. 4.11).

The impact of which is seen in the deterioration of the quality of coastal groundwater.

Coastal aquifers are very sensitive to phenomenon such as sea-level rise, change in climatic conditions, storm surges, shoreline erosion, coastal flooding among others (Barlow 2003). In addition to these, anthropogenic activities also hasten the process of salination in coastal regions. Just like the coastal aquifers, surface water resources

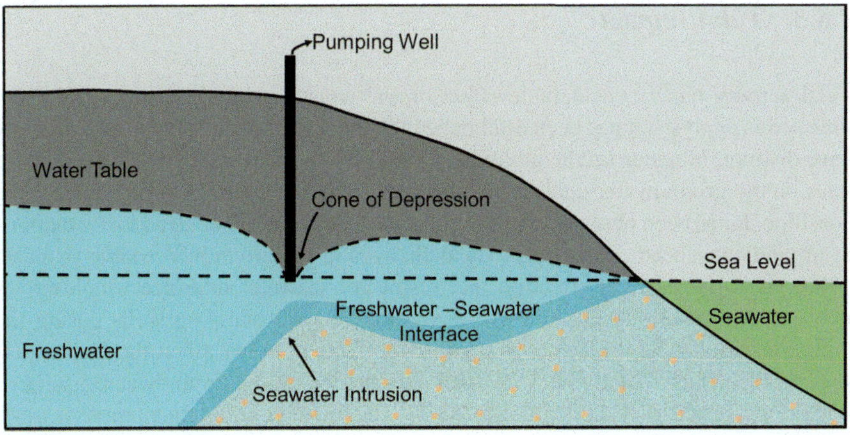

Fig. 4.11 A sketch to illustrate seawater intrusion in coastal aquifer (modified after Prusty and Farooq 2020)

are also affected due to seawater intrusion. Rivers and estuaries permit the natural flow of seawater due to tidal effects making the fresh waters (of this backwater region) more saline. Various factors affecting the coastal aquifers and their effects are summarized in Fig. 4.12 (e.g. Kumar 2006).

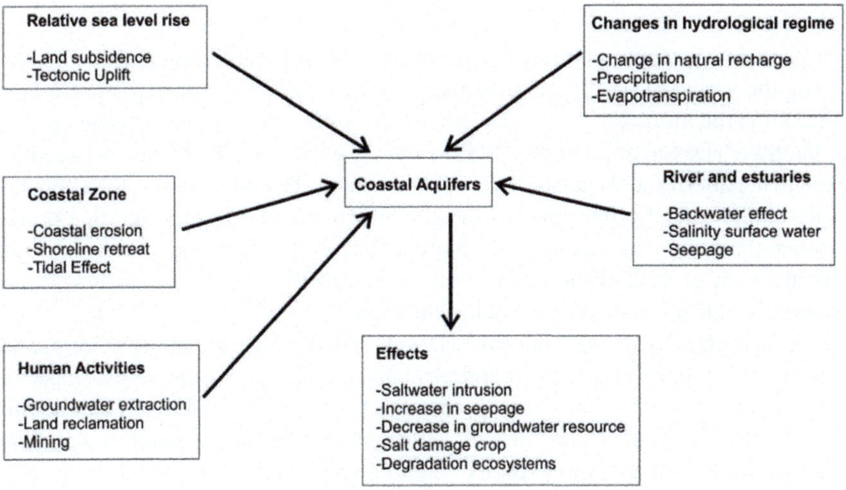

Fig. 4.12 Factors affecting coastal aquifers (after Kumar 2006)

4.6.3 Tidal Impact

Tidal activity has a considerable effect on the coastal groundwater quality. From time series analysis, it has been established that periodic changes in the tidal activity have distinct influence on the groundwater quality (Kim et al. 2005). The effects of tides on the groundwater quality have been traced up to several km inland from the coastline. It has been observed that oscillations in tidal movement lead to fluctuations of groundwater head causing periodical changes in the groundwater table (Nielsen 1990). This has a direct effect on the freshwater-seawater interface which moves upwards and results in the inflow of seawater into the pumping wells during the high tide. Studies have also shown that tidal effects not only affect the freshwater-seawater interface, but also have a distinct influence on the groundwater discharge into the sea (e.g. Urish and McKenna 2004). The impact of tidal fluctuations is more on the watertable of shallow aquifers; the beach-slope plays a significant role in the discharge of groundwater into the sea (Li et al. 2002). Experimental studies, however, have shown something to the contrary, in that, the tide-induced water circulation develops a saltwater plume which in fact reduces the freshwater discharge into the sea thereby controlling the seawater intrusion in the coastal regions (Kuan et al. 2012).

4.6.4 Factors Governing Seawater Intrusion in the State

Local geology, geomorphology of the terrain, the structural features, such as the joints, the cleavage and fracture pattern among others play an important role in controlling the intrusion of seawater in coastal aquifers. Aerial-photo lineament map of the State (Dessai and Peshwa 1982) shows a good correspondence between the lineament pattern and the areas of seawater intrusion. The lithological characteristics of the aquifer are of prime concern and govern the inland flow of water. In a similar manner, the geological history of the aquifer such as the hydraulic gradient, the rate and quantum of groundwater extraction as well as the rate of replenishment also control the rate and extent of saltwater intrusion.

The stabilized (palaeo) dune zone (Dessai 2018) which occurs as a fairly wide zone (0.30–1.9 km) parallel to the coast by and large does not show severe impact of seawater intrusion. The dune zone occurs as a potential perched freshwater aquifer in most parts of the State. The various aeolian and fluvio-marine landforms at many places serve as potential groundwater zones down to a depth of 10 m.

4.7 Confronting the Water Challenge: Global Scenario

Management of water has been a matter of great concern to the global community since water is essential for the very sustenance of all lifeforms/humanity. It is the only commodity that has the capacity to provide food for nearly 6 billion people of the world. Water management and food productivity are thus closely interlinked. It is well known that population grows by geometric ratio, whereas agriculture production increases by arithmetic ratio. In simple words, it means that population growth takes place by multiplication whereas agricultural growth occurs by addition. Hence, growth disparity between the two, will always show an increasing trend. The matter has attained greater relevance primarily due to global warming that seriously affects the spatio-temporal distribution of water. The erratic distribution pattern of precipitation has taken a toll on the biosphere. In such a scenario, to stabilize the biosphere, water is an essential ingredient. The present global requirement of water is of the order of 1700 km^3 and by the year 2050 this requirement would be nearly doubled. In order to achieve the present-day target, one of the measures is to control the surface runoff.

Current approach to water management lays stress mainly on river discharge and groundwater which could be referred to as 'Blue Water'. This limits the option to deal with water scarcity and the vulnerability of water induced by the climate change. An alternative source of water, often given less attention in water management, is the soil water that stems directly from the rainfall. This is known as 'Green Water'. Globally, more than 40% of rainfall goes as surface runoff. The remaining makes up the green water which is a new area of investment for climate adaptation. It forms the basis for food production in rain-fed agricultural systems. It also contributes to the forest cover.

Indian Scenario: In India, the Blue Water makes up about 35% and the remaining 65% is Green Water. Irrigation in India at most utilizes about 11% of the rainfall. Thus, even if the irrigation is doubled, it would be insufficient to meet the food requirement of the country by 2050. The other alternative would be to wisely manage the quantum of Green Water which in other words means greater biomass production, essentially meaning food production from rainwater. This would imply, that to meet the food requirement by 2050 another green revolution would be required.

Water Consumption in Goa Tourism Sector: A comparison of the water uses by the resident population vis-a-vis that by the floating population (which is taken as an approximation of the number of tourists visiting the State) bring out the high rate of water consumption by the latter. The average per capita daily requirement of water of the resident population of the State is estimated to be about 170 l/capita/day in urban areas and 70 l/capita/day in rural areas. This works out to 0.201 Mm^3/year. In some developing countries, water consumption may be as little as 15 l/capita/day; the world average is estimated to be about 60 l/capita/day. About 20 l/capita/day is the minimum quantity of safe water required to realize minimum essential levels of health and hygiene. The approximate quantum of water utilized by the floating population in Goa could be made out from the total room strength of the hotels in the State.

A total of 4961 rooms require 0.68–1.37 Mm^3 of water annually. This requirement is nearly 3.5–6.8 times the water requirement of the resident population. The water requirement for other leisure facilities is over and above that of the room requirement. Nearly six times, the average is the normal requirement for agriculture and industry. However, in recent years it has been observed, that agriculture is an uneconomical activity in the State. The State-agricultural revenue also does not paint a different picture. The industry (other than tourism) requires about 32.46 Mm^3 annually. Most of this demand is met from surface and groundwater storage. Net available water in the State is 2.55 Gm^3. The total groundwater reserves are estimated to be 165 Mm^3 (CGWB 2019). Total abstraction is 53.71 Mm^3. The total quantity is, however, not available for utilization as the storativity as well as the demand for water differs from urban centres to rural settlements. The dependable yield of groundwater is only 28.23 Mm^3 (Anon 2021). Thus, the requirement of water of the floating population is met both from the surface and groundwater. The heavy stress on the resource from the tourism sector is quite evident and has to be maintained at sustainable levels.

4.7.1 Integrated Water Resource Management

The distribution and movement of water in modern times are concerned with the role of human interaction. Although by no means the only motivating force, human need forms a strong basis to understand hydrology. The human interaction is concerned with the quantum of water, e.g. floods, droughts, storage and utilization of surface water, over-extraction of groundwater and the like, and its quality, for example, drinking water, management of aquatic ecosystems, disposal of effluents, and other pollutants and so on. It is concerned with all aspects of the hydrological cycle or more specifically with those that are of direct concern to humans particularly of water consumption. The ***water management includes not only water consumption but protection and management of water for various other activities***, the most important one being the food production, in addition to leisure activities, such as sports, canoeing, boating, fishing and the like. Management of the waste disposal, pollution and water quality, aquatic water system management, flood defence, navigation with the obligation to protect the water environment for future generations and for other species that co-exist with water, are some of the other facets of water management. Water resource management should include all these issues and adapt to the changing views on the requirements of the water management.

An important aspect of water resource management is the involvement of many different sectors of the community in decision-making. This has led to two key concepts in this area: (i) Integrated Water Resource Management (IWRM) and (ii) Integrated Catchment Management (ICM). The concept behind IWRM is also referred to as the 'Dublin Principle'. In 1992, an International Conference on Water and Environment was held in Dublin, Ireland. It was attended by government-nominated experts and those from the non-governmental organizations wherein four guiding principles were generally agreed upon by the stakeholders to manage freshwater resources. These are as follows:

i. Fresh water is a finite and vulnerable resource, essential to sustain life, development and the environment.
ii. Water development and management should be based on a participatory approach, involving users, planners and policy-makers at all levels.
iii. Women play a central part in the provision, management and safeguarding of water.
iv. Water has an economic value in all its competing uses and should be recognized as an economic good.

The four principles enunciated above specify the participatory approach and the economic factor which indicate that water is a commodity that has value to people and that it can be sold at profit in the marketplace/commerce/trade. The emphasis has been on the integration between sectors involved in water resources including the local communities. In other words, it highlights the interconnectedness of hydrology, ecology and land management.

The other concept is the Integrated Catchment Management (ICM) also referred to as Integrated Water Basin Management (IWBM) which in fact is a part of IWRM. Its objective is to promote an integrated approach to the water and land management but with two subtle differences:

i. ICM recognizes a river basin as an unit for understanding and managing water-related biophysical processes in relation to the social, economic and political considerations.
ii. The different management actions are recognized in a spatial context particularly, the significance of the cumulative effect within a catchment area. For example, the cumulative effect of the pollutants discharged at various points, however small, may collectively affect the river environment as a whole.

Fenemor et al. (2006) have defined the ICM concept with three connotations:

i. integration between the local community, science and policy so that the community is involved in the planning and execution of both science and policy, and scientific research is carried out in an environment close-linked into policy requirements and vice versa;
ii. integration between different scientific and technical disciplines to tackle multidimensional problems; and
iii. spatial integration throughout a watershed so that the cumulative impact of different actions can be assessed.

One of the key principles of both these concepts is the community involvement through a participatory approach ensuring that each stakeholder is involved in the resource management.

A closely related concept that is commonly followed in India, United States and some other countries is included under the watershed management.

4.8 Sustainable Watershed Development

The term watershed as explained earlier strictly refers to a catchment area. However, in a broader sense, as used here, the term *watershed management* implies ***prudent use of soil and water resources within a given geographical area so as to enable sustainable production and minimization of the runoff, erosion and floods***. It involves efficient and rational utilization of land and water resources towards maximum production with least hazard/damage to the resource. In terms of the physical components, watershed management is nearly synonymous with soil and water conservation for optimum production. As such all changes in land use, vegetation cover and other structural and non-structural actions undertaken in a watershed to achieve specific watershed management objectives are included under watershed management. Watershed management practices will thus depend on the specificity of the objectives. More relevant objectives of all watershed programmes include:

- increase in infiltration,
- control of damage due to excess runoff and
- management and utilization of runoff for useful purposes.

Several factors influence and govern the watershed operations in an area. They include: (i) physiography, (ii) soil and geology, (iii) vegetation cover, (iv) precipitation, (v) runoff rate, (vi) socio-economic factors and (vii) local population.

During the fifth five-year plan of the nation, the watershed approach got wider acceptance and a variety of programs were included within it. Simultaneously, awareness programmes were conducted to conserve and manage land in the context of the demand from various sectors. A national policy was adopted to use watersheds as a unit to conserve land and water resources.

The State has taken initiative to constitute a watershed management and development authority to oversee the work being undertaken by the State in various watersheds. One of the primary tasks has been of water conservation in addition to the conservation of soil and vegetation—the three principal natural resources essential for the survival of humankind.

4.8.1 Water Conservation in the State: Background

Prior to liberation (1961) from the Portuguese rule, the population of the State was hardly about 5,00,000. There was no domestic water supply except for the two townships of Panaji and Margao. There were no major irrigation projects other than the small water barrage across the River Kushavati at Quepem. Agriculture in the State depended on rainwater and on irrigation carried out by seasonal bunding of streams for the supply of water. Except for the mining industry, which was self-sufficient in terms of water supply required for the mining operations other major industries too were non-existent. Thus, the water requirements of the local population were essentially met from the groundwater supply that invariably showed depletion for a short period during the dry season, in case the pre-monsoon showers were delayed during the month of May.

4.8.2 Post-liberation Scenario: Interlinking of Rivers

After the liberation particularly during the last 50 years, there has been a threefold increase in population (1.45 M; 2011 census) due to the migration from other parts of the country, emigration, and a large floating population of seasonal labour and tourists both local and foreign. Attendant urbanization and industrialization have also contributed to the demographic increase. This has resulted in considerable stress on both the surface and groundwater resources.

Sustainable Water Solutions: Water conservation is the main area of sustainable tourism. A sustainable water system should be able to provide adequate water quality and appropriate water quantity for a given purpose without compromising the future ability to provide this capacity. Thus within the realm of sustainable development, there may not be actual use of water at all, yet may include actions due to which water can be conserved, for instance, the use of waterless toilets and water-free car washes among other initiatives by which there is less stress and hence secure sustainable water supply.

To meet the demand for water, the State in recent years has constructed several small irrigation projects near Tillari, Anjune (Sattari), Salaulim (Sanguem) and Chapoli (Canacona) intended for irrigation and domestic water supply by harnessing a fraction of the runoff. A total of 337 *bandharas* (check dams), including 54 in the Mandovi basin and 37 in the Zuari basin, have been built across major streams of the State's rivers to harness rainwater and to retard the escape of groundwater to the base flow. Interlinking of rivers has been undertaken on a small scale. Zuari River has been linked to Kalay River from the Mandovi watershed via a canal. Likewise, the Chapora River has been linked to Assonora River from the Mandovi basin via the Par tributary. Such initiatives will go a long way not only in conserving and protecting the water resources and floods, for generations to come, but in achieving self-sufficiency in water to meet various societal requirements of drinking water, agriculture, industry, municipal purposes for sewage and sanitation, not to mention navigation and water diversion among others. Further, there are a number of potential sites, upstream of both the Mandovi and Zuari, where dams could be constructed with the least environmental damage. For example, a site at Beedifathar near Mollem was selected for the construction of a dam with the potential to irrigate about 14,000 ha of land. However, despite the water shortage faced by the State in recent years, there is considerable reluctance from the political establishments to significantly contribute to the augmentation of water supply.

Reasons for Caution: It is fairly well known that large dams present serious challenges to sustainability by threatening wildlife habitat, fish migration, water flow and quality and impacting socio-economic conditions of the region. Therefore, an evaluation of the impact assessment of the region in question by a reputed agency should be carried out.

Although agriculture per se has dwindled over the years, the demand for the industrial and domestic consumption of water has indeed increased manifold. The thrust during the decade- 1990–2000 had been primarily to increase productivity by vigorously developing groundwater resources through bore wells and the deepening of existing dug-wells. This has resulted in the steady and at times rapid recession of groundwater levels in several parts of the state. In fact instances of seawater encroachment have been experienced in coastal aquifers mainly due to large-scale withdrawal of groundwater by the hospitality sector that occupies predominantly the coastal belt. Despite the supply of potable water under the urban water supply scheme, the quantity of water required by the coastal belt exceeds the groundwater reserve capability of the aquifer. There is thus, a severe water shortage for about two months in a year, namely April and May, despite the various schemes implemented for irrigation and urban water supply.

Reuse of Used Water: Sustainable water management in cities and towns necessitates the reuse of treated used waters also referred to as 'reclaimed water' which could be utilized for a variety of purposes, such as irrigation, farming, sanitation and the like. Such attempts were initiated with the implementation of geologic, hydrographic and climatologic studies of Panaji town as early as the beginning of last century (de Sa 1908) and were followed in later years by investigating water quality as a sequel to parasitological and epidemiological studies in Portuguese India (de Melo 1914, 1923). These early attempts were followed in later years by the Portuguese medical doctors and scientists, during the 1930s–1940s when the hospital used waters and the municipal wastewaters were treated and reused for sanitation and other purposes. However, in the subsequent years, the thrust has been on locating new water resources rather than on recycling and reuse. Reclaimed water is also a sustainable resource and is an useful initiative to alleviate stress on primary resources such as the surface and groundwater. The use of reclaimed water varies globally, countries such as China and Spain lead in the use of reclaimed water, the latter in particular accounted for nearly 49% of its capacity between 2010 and 2017.

Sewage Treatment: The sewage treatment plants presently operational at Panaji, Margao and Vasco da Gama together treat 0.0783 Mm^3 of water per day. This capacity is likely to increase further to 0.165 Mm^3/day. At present, the treated sewage water is let out in the river system. This water could be judiciously and efficiently utilized for agriculture, industrial use and other related applications or may even be diverted to water-stressed areas.

Change in Land-Use Pattern: Likewise, considering the increased demand for water, erratic and at times below average at others excessive, unequal distribution of annual precipitation, and depleting groundwater resources, the State should be prepared with an alternative strategy for the change of the land-use pattern. In recent years, the use of chemical fertilizers has increased the demand for water, destroying soil-microorganisms and releasing CO_2 and the more damaging N_2O to the atmosphere. The rising levels of CO_2 affect the soil fertility thereby reducing the nutritional levels (protein, zinc, iron) of staple crops, such as wheat, rice among others. In order to protect the food system, CO_2 emissions need to be reduced and new varieties of vegetables, cereals and pulses that consume less water will have to be developed and introduced for cultivation, so that water can be conserved. For example, new varieties of potatoes and tomatoes that consume 40% less water than the conventional varieties have been recently developed in European countries. A situation is evolving/emerging whereby food can no longer be taken for granted. It may be desirable even to change our food habits to some extent in keeping with the land use and availability of water. At the same time, increased awareness of water conservation and sensitization of the local populace towards controlled utilization of water in general, are essential.

4.9 Groundwater Conservation Initiatives

Background: India with a demography of 1.38 billion (2011 Census) is the second most populous country in the world. However, in terms of water potential, it caters to barely 4.0% of the world's freshwater resources. The country often has to face water-stressed conditions due to the changing weather pattern and at times acute water shortage. Safety and security are the main cause of concern due to the inadequate planning for water. Nearly 500 districts of the country are affected by groundwater depletion (CGWB 2019) with as many as 256 districts reported as 'critical' or 'over-exploited' groundwater levels. Hence, an awareness of groundwater conservation attains importance.

Groundwater conservation is concerned with a range of measures aimed at preventing damage to the aquifer and adopting remedial measures to rectify the damage that occurs due to depletion, pollution, water-clogging and the like, as also maintaining the quality and quantity of groundwater that would be beneficial to the economy of the region. Although there are several ways to conservation right from the simplest, checking the leakage from the water-supply pipes, to limiting shower-time, to controlling dish-washers and other washing machines to reduce, reuse and recycle water, some significant initiatives that help the aquifer to retain and regain its optimal water-holding capacity will be discussed below.

Although there is no water scarcity in the State as of today, with the fast-paced development the demand for water is likely to increase in future. Moreover, in view of global warming and the erratic behaviour of the monsoonal pattern, the State should not remain complacent as far as its resource base is concerned. It should be prepared for any eventuality in the event of a climatic catastrophe by carrying out period assessment of the resource, systematic management plan so that the resource is put to efficient and optimal use avoiding wastage and at the same time conserving the natural wealth for posterity by adopting recharge, reuse and recycle practices so that the stress on the resource is considerably minimized and its life is extended to benefit future generations. Hence, a concise account of the various conservation measures, that could be adopted to prevent or overcome water-stressed conditions, is presented.

4.9.1 Conjunctive Use of Water

In keeping with the conservation of water resources, a new concept of '*conjunctive use*' of surface and groundwater resources has emerged. Conjunctive use refers to the use of the surface and groundwater to meet the demand. For example, many farmers in canal-irrigated areas use both the surface water and the groundwater as per their requirement. This is done without any administrative control. This can also be done in a planned manner, wherein conjunctive management comes into play. It refers to the efforts at the basin level to optimize productivity, equity and environmental sustainability by managing both the surface and the groundwater resources. Conjunctive management involves administrative control over the surface water and groundwater. It is more complex, and at times, can be controversial. However, it is very helpful in water-scarce regions and in times of drought, since failure to integrate conjunctive water resources can result in groundwater overexploitation.

A good example of conjunctive use of water is provided by the state of Uttar Pradesh. Here, in recent times, the unused surface water-drains, originally constructed to control waterlogging and floods, were suitably modified by constructing check dams to promote groundwater recharge. The watertable thus rose by nearly 6 m from the previous level of 12 m within a decade.

Conjunctive use of water in Goa has been in existence since ancient times. The rabi crops, also known as the winter crops, of rice, vegetables and even sugarcane, were irrigated partly from groundwater and partly from the seasonal surface water reservoirs created by bunding the streams and rivulets. Such bunds were constructed by the village community by pooling resources. This practice was prevalent in the State up to pre-liberation times, however, in recent years, due to various policies of the State to subsidize agriculture these useful practices have lost relevance.

Dewatering of working mines contributes to significant quantity of water that is generally discharged into the surface drainage system, especially where mining has proceeded below the piezometric level. In a few mines, however, it is utilized for irrigational purposes after assessing the quality of the waters and soils involved. This approach can solve the twin problem of wastewater disposal and the shortage of water for irrigation. The approach here is conjunctive since the mine waters are used to complement inputs from streams or groundwater. These waters can be used for irrigating both virgin areas and the mine-rehabilitated ones or they could be utilized for recharging the aquifers, however, it is essential that the waters pumped out are tested for quality.

During the times of acute water scarcity faced by the State in April/May of 2010, the waters from the abandoned Codli mine pits were pumped into the Assonora reservoirs to maintain the falling water levels of this storage facility.

Conjunctive water management practices can also be attempted in the rapidly urbanized coastal belt and the industrial parks/habitats after proper quality assessment, to ascertain the suitability of these waters for the desired end-use.

4.9.2 Artificial Recharge

Artificial recharge is the infiltration of meteoric water into the subsurface minus the evapotranspiration from the subsurface. It involves the introduction of surface water in basins, furrows, ditches or other facilities where it infiltrates into the soil and moves downward to recharge the groundwater aquifers. Artificial recharge is increasingly used both for short- or long-term underground storage of water as it has several advantages over surface-storage and in water-reuse. Artificial recharge normally requires permeable surface soils and sub-soil lithologies. Where permeable zones are not available, trenches or shafts in the unsaturated zone can be used, or water can be directly injected into the aquifers through wells. However, it is necessary to identify potential recharge areas for the different aquifers and then decide the appropriate measures of artificial recharge. A systematic and continuous monitoring of wells is essential wherever artificial recharge measures are operational.

The State receives more than sufficient rainfall, however, ruggedness of the topography and the limited thickness of the soil apron presents problems in recharging the aquifer and retention of water. The majority of the aquifers from the State are highly permeable, coupled to that the hydraulic gradient is also steep. The groundwater resource thus gets depleted quickly creating water scarcity conditions in the summer months. It is, therefore, necessary to augment recharge to groundwater by constructing water-retention structures, such as percolation tanks, check dams, *bandharas* (weirs) and the like.

Until about mid-nineties there was not much awareness or thrust for augmenting groundwater recharge. In recent times, there have been concerted efforts on the part of the Central Groundwater Board, State Department of Public Works and the Universities to bring about an awareness of both water conservation and artificial recharge. A number of pilot projects and demonstrative artificial recharge schemes, on experimental basis, have been undertaken in the State to sensitize people on the need for artificial recharge to groundwater aquifers. In addition, the State has constructed check dams and *bunds* (embankments) across streams and *nallas* (watercourses) towards checking the surface runoff and to increase the rate of percolation. Scientifically designed artificial water-harnessing structures have been constructed on a pilot scale towards charging the groundwater aquifers, especially during the rainy season. Desilting of old natural and artificial tanks is another method which could be employed to improve the recharge.

Precautionary Measures: The unsaturated zone between the land surface and aquifer must be checked for sufficient permeable zones and absence of pollution. The infiltration rates of the soil must be adequately known. The aquifer should be sufficiently transmissive to avoid excessive build-up of groundwater mounds. Information on these conditions should be gathered from field and laboratory investigations before attempting artificial recharge. Water quality must be evaluated, especially with regard to the formation of clogging layers on basin bottoms or other infiltration surfaces, impact of turbidity on groundwater recharge and discharge (e.g. Aziz et al. 2015) and for potential geochemical reactions in the aquifer. Prior information on all these aspects is essential as the quality of the water in the aquifer and aquifer performance is likely to undergo changes subsequent to the recharge. Instances are noticed where salinity levels of the aquifer waters have been lowered, and turbidity levels have been increased subsequent to the recharge. The latter could be contributed from the incorporation of colloids from the soil during the recharge process (ibid.) With the advancement in technology turbidity measuring sensors are being employed to quantify the turbidity levels in the aquifers.

Pollution from Microplastics: In recent times, pollution of the aquifer from microplastics is an additional cause of concern. The microplastics (refer Sect. 5.2.4.3) comprise particulate plastics 5 mm or less in size as per the National Oceanic and Atmospheric Administration (NOAA). For the purpose of visualization of the size, one can think of something analogous to a sesame seed. The microplastics are present in a variety of products from cosmetics to synthetic clothing to plastic bags and bottles. Many of these materials enter the environment through waste wherefrom some of the chemical additives leach out of the plastics and may enter the soil wherefrom they make their way to the aquifers. The entry to the aquifer could be facilitated often from artificial recharge. Plastic particles can release chemicals into the ground which can leach into groundwater supplies and affect the drinking water quality. Breakdown of plastics could result into new physical and chemical characteristics that can bring about unpredictable effects. Hence, rigorous monitoring of the water quality is imperative followed by constant and strict monitoring of the aquifer with the help of observation wells during the recharge process to avoid untoward damage to the aquifer.

Recharging of aquifers could be undertaken in the mining belt of Goa where large quantities of water are pumped out and released into the surface drainage. These waters could be suitably managed to recharge the aquifer under suitable geological conditions and with the adequate knowledge of local geological conditions.

4.9.3 Rainwater Harvesting

'*Rainwater harvesting*' may be described as collection and storage of rainwater either above or below the ground for later productive use. In India nearly 1300 rainwater harvesting structures with a total storage capacity of 47 Ml provide clean water to over 2,35,000 people from remote rural communities. In recent years, there is an increased realization that the groundwater resource has to be conserved to the extent possible.

Initiatives within the State: There have been attempts mainly from the industry and also from the local populace towards rainwater harvesting. A few artificial rainwater harvesting structures have been constructed particularly by the industry along the coastal belt of the State. All the new construction and housing projects in urban areas are expected to make provision for the construction of water harvesting structures. A few schools in South Goa have constructed roof-top rain harvesting structures. Existing tanks and ponds have been suitably modified and reinforced to augment their water-holding capacity. In some cases, natural structures with potential for water-holding have been modified to increase their water-holding capacity. The waters although are largely used for irrigation and sanitation purposes; in some cases, they have also been utilized for drinking purpose. The water harvesting structures could also be utilized to recharge the groundwater aquifers. Potential areas to undertake these exercises are the abandoned mine pits left after the removal of the ore. Such pits are extensive and have large water storage capacity. They are adequately suitable for recharging the aquifers. The State on its part has taken initiative to popularize cost-effective water augmentation techniques that are suitable to a variety of geological conditions.

4.9.4 Mitigation of Water Crisis

The mitigation of the water crisis needs to be addressed by a multipronged approach. Firstly, there should be a planned and harmonious development so that the future requirements are well anticipated and suitable provisions are made. Various conservation techniques some of them elaborated above need to be put into practice. Reduction of runoff is of considerable importance in a country where over 60% of rainfall, and in some states such as Goa, where reportedly over 80% (Anon 2011), goes as surface runoff. Harnessing this water by means of small dams, *bandharas*, subsurface dykes, percolation ponds and the like will have to be undertaken on a larger scale than attempted hitherto, so that with the fast-paced urbanization and the ever-increasing demand for water is adequately met, consequently reducing the stress imposed on the resource due to overexploitation. The rate of overexploitation particularly in urban, industrial and recreational areas can have adverse effects both on the quality and quantity of this invaluable resource. It is, therefore, imperative that scientific assessment and management of the surface and groundwater resource for sustainable use are undertaken in due earnest as the surface water bodies help in augmenting groundwater resources. Conjunctive use of water should be made compulsory in areas where waters are heavily pumped and discharged into the surface drainage systems such as rivers, without being put to any constructive/effective use. This is especially the case in the mining belt where large quantity of water is continuously pumped out of the working mining pit so that the pit remains operational for mining purposes. Part of this water is utilized for beneficiation in the case of few mines and then left in the natural drainage system. This water could be recycled and could also be put to an alternative use.

Water harnessing by augmentation of recharge to the groundwater and harvesting of rainwater in general, especially in the new urban settlements, could also go a long way in mitigating the water crisis. Reuse of municipal and industrial wastewaters could be yet another method to conserve water and thus contribute towards the mitigation of this precious replenishable resource. Considering the vulnerability of the resource due to the impact of global warming it may be a good idea to initiate water-efficient agricultural practices. Recent studies world over have laid more stress on maximizing the utility of green water. Wise water management practices, so as to conserve both the blue and green water resource in general, would go a long way to come out of water scarcity. A number of attempts are being made to efficiently utilize blue water. The stress now should be to maximize the utility of green water. These practices would build resilience into the system which would simultaneously help to control floods, droughts and dry-spells, so common in modern times.

Although the State is endowed with good groundwater storage and recharge potential, only a small fraction of the total groundwater resource is being utilized for the benefit of the populace. There is a need for better evaluation of the resource taking into account potential aquifer storage. The management of the resource too should place greater emphasis on the prudent utilization of groundwater so as to conserve the resource for posterity, as some of the estimates even after three successive assessments do not show real-time estimates of change in the aquifer storage in the different aquifer systems. Hence, caution needs to be exercised in designing local groundwater management initiatives. The State has to invest more on the conservation and revitalization of the resource. Conjunctive use of surface and groundwater should be one of the critical management options towards efficient utilization of both these resources and this should be coupled with regulating the land-use practices. With the rapid pace of urban- and rural-development and significant stress from the tourism industry water in general, and groundwater in particular, is likely to play a bigger and more crucial role in the developmental activities of the State. Considering the criticality of a depleting resource the need to protect and conserve it in land-use planning is vital and cannot be overemphasized for ensuring the sustainability of the fundamental component of the national economy.

References

Adyalkar PG (1985) Hydrology of Goa with particular reference to the lateritised midlands and the coastal strip. Proc. Seminar Earth Resources for Goa's Development, Geol. Surv. Ind., pp 484–494

Anon (2011) Goa Development Report, Planning Commission, Government of India, New Delhi, Academic Foundation, New Delhi, pp 1–250

Anon (2021) The Goa State Water Policy, Official Gazette, Government of Goa, Series 1, No. 23, pp 1213–1244

Aziz A, Yusuf H, Faisal Z, Suradi M (2015) Water turbidity impact on discharge decrease of groundwater recharge in recharge reservoir. 5th International Conference of Euro Asia Civil Engineering Forum. Proc. Engineering 125:199–206

Barlow PM (2003) Groundwater in freshwater-saltwater environments of the Atlantic coast. US Department of the Interior, US Geological Survey, Reston, VA. http://pubs.water.usgs.gov

Central Groundwater Board (1984) Groundwater estimation methodology. Report Groundwater Estimation Committee, Ministry of Irrigation, Government of India, pp 1–49

Central Groundwater Board (2013) Ground water information booklet, North Goa district, Goa State, Southwestern Region, Bengaluru, pp 1–22

Central Groundwater Board (2019) Aquifer mapping and management of groundwater resources, Goa State, Southwestern Region, Bengaluru, pp 1–24

Chaulya SK, Chakraborty MK, Ahmad M, Singh KKK, Singh RS, Tewary BK, Gupta PK (2000) Water resource accounting for an iron ore mining area in India. Environ Earth Sci 39:1155–1162

Choudri BS, Chachadi AG (2006) Status of groundwater recharges and availability in the mining areas of North Goa. https://www.researchgate.net/publication/258100830

De Melo F (1914) Conferencia Sanitaria de Lucknow. Relatorio Apresentado ao Governo da India Portuguesa. Imprensa National, Nova Goa, 335 pp

De Melo F (1923) Onze anos de investigacoes laboratoriais em Nova Goa. Imprensa National India Poertuguesa, Ciclo 4:34

De Sa I (1908) Primeira tratamos de Pangim acerca das sua historia, geographia, geologia, hydrographia, climatologia, demographia e nosologia na segunda acerca, Typographia Arthur & Viegas, Nova Goa, pp 1–104

Dessai AG (2018) Geology and mineral resources of Goa. New Delhi Publishers, New Delhi, 323 pp

Dessai AG, Peshwa VV (1982) Manganese mineralization in the tropical forest area of Goa, India. In: Laming DJC, Gibbs AK (eds) Hidden wealth: mineral exploration techniques in tropical forest areas, Proc. Seminar of the Association of Geoscientists for International Development, Caracas, Venezuela, pp 170–175

De Wiest RJM (1965) Geohydrology. Wiley, New York, p 366

Fenemor A, Phillips CJ, Allen WJ, Young RG (2006) Integrated catchment management-interweaving social process and science knowledge. NZ J Mar Freshwat Res 45:313–331

Hamed Y, Hadji R, Zigmi B, Baali F, Gayar El (2018) Climate impact on surface and groundwater in North Africa: a global synthesis of findings and recommendations. Euro-Mediterranean J Environ Intergr 3. https://doi.org/10.1007/s41207-018-0067-8

Kim J-H, Lee J, Cheong T-J, Kim R-H, Koh D-C, Ryu J-S, Chang H-W (2005) Use of time series analysis for the identification of tidal effects on groundwater in the coastal area of Kimje, Korea. J Hydrol 300:188–198

Kuan WK, Jin G, Xin P, Robinson C, Gibbes B, Li L (2012) Tidal influence on seawater intrusion in unconfined coastal aquifers. Water Resour Res 48:2501–2502

Kumar CP (2006) Management of groundwater in salt water ingress coastal aquifers. Groundwater Model Manag 8:540–560

Li L, Dong P, Barry DA (2002) Tide-induced watertable fluctuations in coastal aquifers bounded by rhythmic shorelines. J Hydraul Eng 128:925–933

Nielsen P (1990) Tidal dynamics of the water table in beaches. Water Resour Res 26:2127–2134

Prusty P, Farooqui SH (2020) Seawater intrusion in coastal aquifers of India—a review. HydroResearch 3:61–74

Shiklomanov I (1993) World fresh water resources, In: Gleick PH (ed) Water in crisis: a guide to the world's fresh water resources. US Geological Survey, Water Science School. http://www.usgs.gov/special-topic/water-science-school

Tideman EM (1996) Watershed management, guidelines for Indian conditions. Omega Scientific Publishers, New Delhi, 372 pp. http://books.google.com

Todd DK (1980) Groundwater hydrology. Wiley, New York, 535 pp

Todd DK, Mays LW (2005) Groundwater hydrology. Wiley, Hoboken, 635 pp

Urish DW, Mckenna TE (2004) Tidal effects on groundwater discharge through a sandy marine beach. Groundwater 42:971–982

US Geological Survey (2019) The distribution of water on, in, and above the Earth. http://www.usgs.gov/special-topic/water-science-school

Van Camp M, Mtoni Y, Mjemah IC, Bakundukize C, Walraevans K (2014) Investigating seawater intrusion due to groundwater pumping with schematic model simulations: the example of Dar es Salaam coastal aquifer in Tanzania. J African Earth Sci 96:71–78

Chapter 5
Environmental Assessment

Abstract Mining and to a lesser extent agriculture were the two basic industries that shaped Goa's economy post-World War II. However, since early eighties tourism has also emerged as a predominant industry to reckon with.

Goa has been a favourite tourist destination for at least three decades. The floating population in the State for the major part of the year is nearly double the resident population. The State has also witnessed open cast mining and rapid industrialization for more than seven decades. All these anthropogenic activities have been major contributors for the generation of waste of varied types that has impacted the physical environment and ecology of the State in diverse ways.

The pollution of rivers and standing water bodies has impacted the aquatic life, affecting the spawning patterns, reproduction and growth rates upsetting population levels. Good quality agricultural lands that occur as tidal flats along the estuarine sections of rivers with fairly luxuriant mangroves are the productive and ecologically sensitive areas, that are the most threatened due to soil erosion, poor farming practices, overfishing, drainage and sewage discharge, release of industrial effluents, reclamation, mineral exploitation and several other anthropogenic activities.

The pollution of air is yet another cause for distress due to potential impact on human health, the ecosystem and climate.

The effluents from the industrial estates, metallurgical units and fertilizer plant, solid waste including biodegradable waste, bio-medical waste, electronic waste, construction and demolition waste, plastic waste are the major contributors of pollutants to the land and river systems. The threat to the coastal marine ecosystem from ballast waters is an additional cause of concern that could be a menace to public health and marine environment, in addition to other damaging effects.

Keywords Particulate matter · Biodiversity · Wetlands · Ballast waters · Microplastics · Groundwater pollution

A. G. Dessai, *Environment, Resources and Sustainable Tourism*,
Advances in Geographical and Environmental Sciences,
https://doi.org/10.1007/978-981-99-1843-0_5

5.1 Introduction

Earth is a dynamic planet; its dynamism is manifest by a variety of topographical features, landforms, mountains, gorges, oceans, rivers, lakes, rocks, minerals, fossils, soils among others which make up the geodiversity of the earth. In fact geodiversity is the non-living elemental foundation of nature and all its biodiversity and hence of human diversity. Whereas, the geomorphological features represent a variety of habitats, potentially inhabited by different organisms (Gray 2005; Gordon and Baron 2013), the fossil record is a part of geodiversity that testifies the evolution of biota in the geological past (Shikazono 2012), playing an important role in formulating strategies for biological conservation. The palaeobiological conservation integrates palaeontological and ecological information that can be used as evidence to understand current environmental modifications and alterations. As such geodiversity is one of the precious resources of the planet that serves as the bedrock of natural and cultural diversity and is closely linked to biodiversity (Tukiainen 2019). Therefore, it follows that just as we strive to preserve biodiversity, we should be equally driven towards the protection and preservation of the geodiversity, that broadly includes the physical environment.

Environmental issues are primarily focused on humans and relate to their life and life support systems. Environment and development are complementary to each other. It is important that the development be in consonance with the protection of the environment. Sustainable development finds its roots in this context.

No account, especially the one that deals with the exploitation and mining of natural resources is complete, unless it deals with the environmental implications and effects, that are observed in the region, and that are likely to threaten it in future. They directly influence the quality of life and well-being of the people, and communities involved or affected. Moreover, Goa has been subjected to sustained mining activity that has been in existence for a period of over six decades. It is one of the famous tourist destinations of the world since the last forty years. As the areal spread of the State is limited, the pressures on the environment are tremendous, both because of the floating tourist population and migrations from other parts of the country in search of better prospects. In such a region, environmental concerns and the impact of environmental degradation cannot be ignored. In fact, it is one of the ideal areas to study the brunt of mining and other developmental activities on the environment and the latter's response to these shocks.

5.2 Environmental Concerns

Goa, as many other regions of the country, has shown great awareness and concern to the environmental issues. The State has been a front-runner in raising environmental issues and voicing them not only at the local and national level, but even internationally, thanks to the Indian (Goan) diaspora in other parts of the world. It is worth analysing why Goa has been so active in raising environmental concerns in comparison with the other states in India. Some of the prime factors among others that come to mind are: (i) its small size; (ii) high literacy rate (87–88%) and (iii) high per capita GDP which is nearly 2.5 times that of the country as a whole. The size of the State has been the prime factor (population to land ratio is small) that has affected and influenced every activity both, beneficial to the State or otherwise. Being small and subjected to rapid increase in population (largely migratory) in recent years, primarily due to the boost to tourism, there is considerable stress on the availability of land for any activity to be undertaken, may it be for harnessing of natural resource or for any other developmental purpose. With the result, some constituent of the environment is invariably adversely affected.

As the State is smaller than even a district of other states in India, it is easy to politicize the issue either with the help of the party in power or with the help of the opposition. Past experience has shown that at times such environmental alarms have been false and even detrimental to the development of the State. Financial support required to maintain and sustain the tempo of agitations for the issue in question, is usually made available through generous donations at times from individuals, communities and corporate houses. The Goan diaspora abroad also plays a significant role in mobilizing support both societal and financial.

5.2.1 Control on the Components of the Environment

Every activity in the State, whether big or small, has been raised as an environmental concern at some time or the other. Surprisingly, the mining activity, despite being extensive and most detrimental to the environment in several ways, was possibly the only activity that had escaped the wrath of environmental concerns (until the Hon. Supreme Court of India imposed ban on mining on 9th September 2012) despite its large spatial and long temporal extent. The reasons appear to be obvious.

Most mining activity, since the initial times is carried out by locals who have consolidated their base over time into large business houses. They have the ability to quell/quieten any environmental discomfiture by utilizing a small fraction of their net earnings to silence individuals, communities, local self-governing bodies (gram panchayats), politicians or even the government's environmental regulatory and monitoring machinery. In the case of affected communities, this is usually accomplished by providing financial assistance to start a business, scholarships for students from the affected families and compensation for the loss of crop or agricultural land, as the case may be.

The other aspect is the media, especially the print media and in recent times the electronic media. Both are largely controlled by the industrial houses. It is solely this multipronged approach to handle situations that had insulated the mining industry from being targeted for environmental issues. They range from dust pollution, to forest degradation, to recession of groundwater table, siltation on farm/agricultural lands, pollution of surface water and groundwater bodies, accidental deaths during various stages of mine operations among others.

5.2.2 Societal Change Leading to Environmental Issues

Prior to the mining boom of 1948 and the period thereafter, the State economy was essentially agrarian. The first transition towards a change occurred in the period post-1948. The economy during this period was mainly controlled by mining, agriculture and, to some extent, fishing. Progressively mining attained a major role in the control of economy of the State. This continued albeit with some variations until the seventies. Since the time Goa hosted the retreat of Commonwealth Heads of Governments Meeting (CHOGM) in 1983, tourism received a boost and Goa was recognized as an international tourist destination. During the period that followed, the State accorded industrial status to tourism. In the ensuing period, the promotion to tourism and boost to education have charted the economy on an altogether different course. Today, the economy of the State is largely dependent on tourism and to a lesser extent on mining, construction and fisheries. Agriculture plays an insignificant role, despite huge investments by the State to improve and augment the infrastructure required by the agriculture industry, needless to mention, the various types of subsidies provided to the agriculture sector.

5.2.3 Environmental Challenges

Deforestation: In the open cast mining, the first serious casualty is that of vegetation cover. It is required to be cleared from the actual site of the mine, the site where overburden is stacked, and from the clearances for the construction of communication network for the servicing of mining operations. The forest covered area of the State is 1,424 km^2 which is approximately 38% of the total area of the State (Fig. 5.1). Of this, about 62% of the area is protected by one National Park and six wild life sanctuaries covering a total area of 755 km^2. The N-S Patradevi to Pollem six/eight lane highway passes through 23 ha of protected and reserved forest. Similarly, the E-W Goa to Belgavi fourlane highway is likely to impact 41 ha of protected forest. The construction of the third Mandovi bridge, the 'Atal Setu', destroyed 2.6 ha of forest. The laying of second railway track between Kulem and Vasco da Gama is likely to impact 17.51 ha of forest of which 14.4 ha in protected area, and the section between Kulem and Castlerock will affect 98.17 ha of thickly forested area of the Western Ghats. A high-level working group report (2014) on Western Ghats led by Kasturirangan identified 99 villages from Goa's Western Ghats as ecologically sensitive areas that should be protected from all anthropogenic activities. The National Green Tribunal also has passed an order to maintain status quo in all these 99 villages.

The threat to forest due to rapid pace of anthropogenic activities is a matter of great concern, especially for a tiny state such as Goa.

In addition the State has witnessed large-scale forest degradation, nearly 2,000 ha due to mining. The figures are placed at 50,000 ha by Valdiya (1997). In more recent times, twenty-six mining leases located within or at the periphery of 1245 ha of forest

Fig. 5.1 Map depicting the distribution of reserved forests and mining lease areas from Goa (after Pascual et al. 2013)

land led to the destruction of 508 ha of it. Impact on forest degradation is high in Bicholim, Sattari, Ponda, Quepem, Sanguem and Canacona *talukas*. The concern of forest degradation is more, since this region of the Western Ghats has high level of biodiversity and endemism, and hence forms a part of this UNESCO-WHC identified biodiversity hotspot.

Threat to Forest Biodiversity: The forests that are cleared for mining operations are home to a large number of organisms which lose their habitat due to the indiscriminate clearing of the forests. This often threatens the survival of a number of animal species. The cutting down of trees is a threat to the forest biodiversity, i.e. plants, trees, birds, insects and other animals that dwell in the forests. Of the over 1500 species of plants identified, at least around 100 are endemic. Similarly at least some of the 325 globally threatened floral and faunal species, found in the Western Ghats, dwell in these forests.

Loss of Soil Cover: Attendant to the above, is the loss to the soil cover due to increased soil erosion which has led to enhanced sedimentation in the streams and rivers, and consequent loss of soil fertility. These have left soreful scars of mining on this otherwise lush green countryside. Large quantities of overburden pushed downslope in the valleys have buried streams and springs. A distressing example of this is seen in the case of Aravalem waterfall which has no base flow except for the rainy season, as mentioned earlier, due to withdrawal of large quantity of water which has led to the recession of the watertable. Such situations are also encountered in other watersheds within the mining belt.

Reshaping of Landscape: The movement of the overburden is responsible for reshaping of the landscape, simultaneously affecting the groundwater aquifers. Large volumes of debris, nearly thrice the volume of the resource mined, is normally stacked as rejects. Storm water runoff from mine dumps and tailing points lead to the formation of inorganic blanket of clay and silt fraction over top soil rendering it uncultivable. The particulate matter is also redistributed downstream on the valley-floor as rain-charged debris and mud-flows, chocking irrigation channels and streams, and rendering vast areas of fertile agricultural land fallow.

Agriculture Distress-Tidal Flats: Good quality agricultural lands occur as tidal flats (Fig. 5.2) such as those at Carambolim, Chorao, Agacaim, Cortalim, Chinchinim, Cavelossim, Bati, Tuem and many other places, along the estuarine sections of rivers. These areas are lower than the high tide level and hence are regularly inundated by the tidal waters. Most of these wetlands are natural, but some are man-made. These are locally known as '***khazan***' or '***cantor***' lands (refer Sect. 7.4.1). They vary in width from tens to hundreds of metres, extend for a few km along river banks and occupy an area of about 17,200 ha. They have resulted from the mixing between sediment-laden fresh waters and saline ones that result in flocculation and settling of clays and silt. Fairly luxuriant mangroves (salt tolerant shrubs or small trees also called halophytes that grow in coastal, saline or brackish, anoxic, waterlogged environments) covered these stretches of land, however, with increasing urbanization and settlements, today solely the relicts of mangroves appear to fringe these estuarine stretches (e.g. Brown 1997). They are particularly well developed along the estuaries of Mandovi and Zuari. Extensive tidal flats are formed at Carambolim, along the

Fig. 5.2 Extensive tidal flat used for agriculture, on the southern bank of the river Chapora at Tuem (after Dessai 2018)

banks of 'Cumbarjua Canal' distributary, in the vicinity of Panaji along the banks of Mandovi river, Agacaim-Cortalim along the Zuari, those along the banks of river Sal and along estuarine sections of all other rivers. These, productive and ecologically sensitive areas, in general, are the most threatened. They are subjected to degradation due to a variety of factors such as soil erosion, poor farming practices, overfishing, drainage and sewage discharge, release of industrial effluents, reclamation, mineral exploitation and several other anthropogenic activities.

Salt Pan Ecosystem: Traditionally, the tidal flats are used for agriculture, paddy (rice) cultivation, by regulating the entry of tidal waters. In the vicinity of Panaji, and elsewhere too, they are used as salt pans, for the extraction of common salt, after natural desiccation. Many tidal flats have been used traditionally for pisciculture, particularly along the banks of river Sal. In recent years, they have been modified especially for prawn farming. In several instances, the *khazans* have been impacted by mining and quite often are reclaimed by illegal construction activity. They have been rendered unproductive due to deposition of particulate matter (silt) from mining rejects, making them unsuitable for agriculture and other specific uses.

Destruction of Biodiversity: The marshes and swamps are threatened due to pollution and destruction of biodiversity in these ecologically sensitive areas, particularly due to development pressures along waterfronts and in the vicinity of towns. The abundant hydrophytic vegetation of these areas sustains large planktonic and benthic macro-invertebrates and abundant fish population. The ambience being rich in nutrients, it supports several types of biotic communities with a complex food chain. These wetlands are ecologically important as they provide ideal conditions to shelter a variety of biota, especially birds (163 species are recorded of which 52 species are migratory), which depend on them for feeding and breeding.

Biogeomorphology of Wetlands: In the last couple of decades, biogeomorphology of these wetlands has emerged as an established research field that examines the interrelationship between organisms (microorganisms, animals and even plants) and geomorphic processes, both marine and terrestrial. Such interactions are crucial to development and understanding of environments such as salt marshes, mangroves and other type of coastal wetlands. In such areas, three types of processes are operational namely, bioerosion (erosion of substrate by organisms, for example through bioturbation), bioprotection (protection of substrate from erosion by the presence of organisms such as coral reefs) and bioconstruction (physical construction of biological structures by accumulation of carbonate sediments). Interactions between marine biota and natural processes are very significant to shoreline stability, especially in soft sediment environment. Benthic and planktonic biota, and shellfish, filter, package and bind sediment together, especially in tidal regions. This goes a long way to control the turbidity by solidifying and protecting loose, soft sediments and thus, allowing for more colonization by other organisms.

Impact from Mine Discharge Waters: It is estimated that 30 Mt of rejects have scattered over 10,000 ha of paddy fields (Fig. 5.3) and coconut groves, as against the annual average export of 12 Mt of iron ore (J. D' Souza, in *Times of India*, dtd. 13.02.1984). The suspended particulate matter in the mine discharge waters comprises clay (<2 μm) mineral fraction which flocculates and serves as a major threat to sustainability of fertile agricultural lands, particularly the paddy fields. This is in addition to siltation from surface run-off. During 2007–2012 period, however, the quantum of rejects was a whooping 700 Mt considering ore to reject ratio of 1:3. This has been the most serious complaint of the communities in the villages adjoining the mines. Over half of the 300 odd mining leases, are located close to the water bodies. Most of the mines are located in the close proximity of Mandovi and Zuari rivers. Several of the 182 mining leases from South Goa occur within one km of the Salaulim dam which provides drinking water to a population of over 6 lakhs. Part of the particulate matter is deposited in standing water bodies causing siltation of ponds and *lakes*/tanks, and reservoirs downstream. One of the worst examples of siltation is seen at Mayem *lake*. Surface run-off of rock debris, although largely non-toxic, can be harmful to the growth of vegetation in the surrounding areas and may have harmful effects on human health (refer Sect. 6.3).

Erosion of Mine Waste: The mineral industry is a major source of waste generation as spoils, tailings, slag and other rock- and mineral-waste (e.g. Montgomery 2006). Excessive mining leads to aggravation of erosion and can induce landslides and rockfalls. Eroded mine waste is a significant pollutant of surface water bodies. It is a matter of serious concern due to the mine-wash waters and leakage of chemicals on to the land. Discharge of waters from wet mine pits and effluents from 28 beneficiation plants contribute to the concentration of suspended inorganic solids to the water bodies.

Waters released from Ore Beneficiation: The quantum of water released from ore processing plants is considerably large, as one ton of ore requires 1.5 m^3 of water for beneficiation. Prior to the Chinese boom which began in 2004, almost entire production was required to be beneficiated by washing. If the total ore production

Fig. 5.3 Rice (paddy) fields rendered uncultivable due to the deposition of particulate matter washed down from the mine dumps (courtesy: R. Gauns)

during 2003–04 was 22.09 Mt (Table 2.2), it required 33.14 Mm^3 of water. This quantum of water is almost equivalent to half of the annual water requirement of South Goa (appox. 59.40 Mm^3) presuming the population is 0.6 million. Although part of the water is reused in the beneficiation process, ultimately the waters are released and thereby affect the natural system.

5.2.4 Threat to Aquatic Environments

The aquatic environments, apart from the standing water bodies, in general, are endangered by a twin threat, one, from particulate matter, chemical pollutants, oils, greases, and the like, largely contributed to the estuarine environments from the hinterland barge traffic and associated activities and, two, from offshore pollutants from potential bioinvaders-exotic, introduced and 'alien' species in marine environments from the inbound ship-traffic.

Impact on Aquatic Life: Turbidity of waters influences light penetration, consequently energy is released as heat, waters get warmer, thereby decreasing the oxygen content. Photosynthetic activity of plants and algae is reduced, lowering the quantity of dissolved oxygen in waters. Increase in temperature and decrease in the oxygen content influence the survival of aquatic organisms. Turbidity also limits the visibility of aquatic organisms impairing the ability to feed. Turbid waters also tend to clog the gills of the aquatic animals leading to suffocation of benthic flora and fauna.

Incorporation of toxic constituents is harmful to aquatic life. It severely impairs the fish habitat influencing the spawning patterns, reproduction and growth rates, affecting population levels. Dissolved iron has an adverse qualitative and quantitative effect on phytoplankton population.

Effect on Soil Fertility: Since, some of the chemicals are poisonous, they make the soil unsuitable for the growth of plants. In addition, the organisms that live in the soil find the polluted environment unsuitable for their survival. In certain cases the mine waters may contain heavy metal concentrations beyond permissible limits for potable water. These waters contaminate the groundwater aquifers. Withdrawal of large quantities of water from wet mine pits changes the water regimen affecting the growth of vegetation and agriculture in the surrounding areas. Thus, changes in aquatic biodiversity have a cascading effect on the species dependent on it.

Previously buried metal sulphides are exposed due to mining operations, although sulphide-rich ore bodies are of restricted occurrence. They thus come in contact with atmospheric oxygen and get converted to sulphuric acid and metal oxides. They get mixed with local waterways and eventually contaminate rivers with heavy metals.

Estuarine Environments- Background: Rivers are the lifeline of the State. In addition to providing food and water, they serve as useful means of communication and transport. It is solely due to the well-developed river network which has provided a cheap mode of transport, the iron ores from the State, that are otherwise lean as compared to those from the rest of the country, have acquired importance, and are sought after the world over.

The major rivers of the State namely the Mandovi and Zuari, which flow through the mining belt, are the most impacted. A large number of barges, 396 to be specific, ply the estuarine stretch of the rivers to transport the mineral wealth from the hinterland loading points to the nearest port located at Mormugão, a distance of hardly 25–30 km. They make about 15,000 trips per year from the loading point to the harbour and back in Mandovi, and about 18,000 trips in Zuari. This magnitude of traffic contributes to re-suspension of bottom sediments due to propeller movement increasing the turbidity of waters. It also adversely affects the dissolved oxygen levels, in addition to contributing oils and greases to the waters. If it is presumed that one single barge discharges a minimum of one litre of oil per day, the total contribution amounts to about 400 l per day which work out to nearly 0.14 Ml of oil per year. This quantity of oil is being added to the river waters on a regular basis, for the last at least 60 years, the maximum contributions, however, being during the last three decades, when the export of ore was at its peak. Import of coal and its hinterland transport by road, rail and through waterways is another potential threat that the State may have to witness in the foreseeable future.

Several barge maintenance workshops and spillways are constructed along the banks of the Mandovi and Zuari including the Cumbarjua canal. These maintenance workshops also contribute to the waters, spent-oils and other lubricants, whose levels vary from 0.2–0.9 mg/l. The land for the construction of the workshops and spillways has been procured at the cost of clearance of mangroves and *khazan* lands (Fig. 5.4) which have been shrinking in course of time.

Fig. 5.4 Endangered Zuari river ecosystem with several barge maintenance workshops seen in the background

In addition, several river cruises (6–8 vessels) and floating stationary casinos, so-called 'offshore casinos' (4 vessels) operate round the year for consumeristic services in the Mandovi river within a limited stretch of about two km, extending from old secretariat (Palacio do Idalcão) to Mira Mar. This stretch of the river waters shows heavy bacterial contamination, (total coliform ranges from 450–35,000 MPN/100 ml (MPN: Most Probable Number); and faecal coliform from 450–24,000 MPN/100 ml, BOD (Biochemical Oxygen Demand): 1.7 mg/l, COD (Chemical Oxygen Demand): 63.7 mg/l) with dissolved oxygen varying between 4.9-7.1 mg/l. The low to critical values of dissolved oxygen in some areas suggest polluted waters, despite the fact that the casino vessels reportedly dispose the sewage through night soil tankers, for treatment at the sewage treatment plant at Tonca. Additionally, treated sewage waters from the sewage treatment plant, are also discharged into the river. The BOD of the waters at the discharge area varies from 42–264 mg/l, COD from 85–400 mg/l and suspended solids from 18–30 mg/l for the period January to March 2014 (Anon 2014). These values are suggestive of microbial activity related to sewage. In contrast, a few km upstream from the old secretariat, at Marcel, the levels of BOD (1.3 mg/l), COD (11.6 mg/l), total coliform (3412.3 MPN/100 ml) and faecal coliform (2153.3 MPN/100 ml) are much lower than those downstream. The high values of coliform in the lower reaches of the river likely suggest sewage contamination and faecal remains.

Regular river transport, at several locations across most rivers, meant to ferry commuters, also contributes to river pollution. Besides, over 863 (out of 1157 registered in the State) mechanized fishing trawlers operate within the coastal stretch, in addition to 1418 motorized canoes and 781 unmotorized ones. This magnitude of activity and density of vehicular traffic in the coastal stretch of not more than 30 km imposes considerable stress on the estuarine environments. It poses a threat to the estuarine biota, largely due to oil spills, dumping of untreated sewage, industrial effluents and demand-oriented mechanized fishing. The carrying capacity of the rivers needs to be estimated, assessed and evaluated in due earnest.

Domestic Sewage: Apart from oil spills, the other major pollutant is untreated domestic sewage that is discharged into the waters by the inhabitants. Industrial effluents and domestic sewage have considerably damaged these fragile estuarine habitats by contributing nearly 30% to BOD and suspended solids. They have altered the natural productivity and energy flux rates of these estuarine environments which are considered to be the nurseries for the commercially important fishes and invertebrates. A pilot survey of Sal river, to study heavy metal contamination concluded, that pollution is primarily due to organic contaminants from municipal waste and heavy metals such as Hg and Pb from dumped waste of a former treatment facility (Anon 2016). Except for the major towns of Panaji, Margao and Vasco da Gama sewage treatment facilities in other towns are not yet available and this is a major cause of concern.

Declining Fish Production: In recent years, fairly large quantity of 'dry ice' is contributed to the riverine ecosystems by the pisciculture industry. Dry ice also referred to as 'cardice' by British chemists, is the solid form of carbon dioxide. It is primarily used as a cooling agent as it sublimes at −78.5 °C at atmospheric pressures. It is used for preserving different foods, ice creams among others. Unused dry ice is dumped into the waters mainly by the pisciculture industry, thereby increasing the CO_2 content of the waters.

Observations during the past decade (Rivonkar, per. comm.) have shown that stocks of several species such as the Spadenose shark-*Scoliodon laticaudus*, Banded dogfish-*Chiloscyllium griseum*, Dwarf whipray-*Himantura walga*, *Arius maculatus*, Tiger prawn-*Penaeus monodon*, bivalve clams-*Paphiamalabarica, -Meretrix casta*, mussels-*Pernaviridis*, Window-pane oyster-*Placuna placenta,* oyster-*Crassostrea madrasensis* have declined. Some of these species, particularly *Meretrix casta*, an edible clam, used to be quite common in the sandy estuarine environments of Sal and Zuari rivers among others, until a few years back. Its occurrence and distribution has become restricted and the abundance has decreased considerably in recent years, possibly due to the change in its habitat, largely the result of human interference. It is a species that prefers sandy habitat. Contribution of large quantities of silt and clay fraction to the estuaries and modification in salinity conditions, primarily due to change in land-use pattern (urbanization, mining and similar activities) has possibly led to the decline. The reduction in the productivity of these resources indirectly imparts stress on the socio-economic status of the coastal population. Recognizing the threat to these resources, the State has now issued guidelines for sustainable utilization, management and conservation of the bio-resources (*Navhind Times*, dtd. 24.9.2022).

Marine Ecosystem-Ballast Waters: The term 'ballast' is used for any solid or liquid material, other than cargo, which is brought on board a vessel to increase the draft (the minimum depth to which a vessel's hull can go without hitting the bottom) and to provide stability to the ship or vessel. In the past, the ballast used in ships were solid rocks (quarried stones) or sand, however, in modern times waters are employed for the purpose. Thus, ballast water is the water that is carried in a ship or vessel to improve its stability, manoeuvrability and attain balance. A compartment within a ship is specially designed for the purpose and it is called a ballast tank. The water is usually taken up when the cargo is unloaded and discharged at the next port-of-call before the loading of fresh cargo. It is also taken up when ship needs extra stability particularly in foul weather. The discharge of this water often alters the aquatic environment of the region, especially in areas of heavy maritime traffic and in the vicinity of harbours or ports and this is a cause of concern.

The water taken up for ballasting also carries with it the biota especially the microorganisms that live in it. These are often the characteristic of the region where the waters are taken. When the ballast waters are discharged at the next port-of-call, they form one of the important sources for the transfer of biota to the new environment. The water-borne organisms, which are alien and non-indigenous to the place of discharge, can become marine pests and can disturb the biodiversity of the region.

Pollutants: The discharged waters may contain a variety of harmful substances, including oil contaminants, non-native marine biota and disease-causing organisms such as bacteria and pathogens from sewage-contaminated water. If the introduced species are aggressive and fast-reproducing, they may become invasive, out-competing local flora and fauna. Therefore, the introduction of non-native species is considered to be one of the five major threats to marine biodiversity identified in the Convention of Biodiversity, 1992. The introduction of non-native species from ballast water, in addition to other sources, is a matter of grave concern and is a potentially serious, but highly unpredictable problem, in all coastal marine ecosystems. For instance, it is estimated that there are ninety, non-native, marine species in British waters, of which fifty eight are established. These have been introduced by shipping, both in ballast waters and imported consignments of cultured species (Minchin et al. 2013).

Disruption to Ecosystem: The effects of introducing new biota can be almost imperceptible, or dominating, so as to displace the native communities. They often disrupt ecosystems by monotypic colonization affecting production of fisheries and aquaculture as well as the economy and livelihood of communities dependent on the invaded area's biodiversity. They even damage harbours, ships and water treatment and power plants. Instances of introduced, non-native organisms include the European zebra mussel (*Dreissena polymorpha*), native of lakes from Russia and Ukrain, into the North American lakes, causing billions of dollars' worth of damage due to fouling (living or non-living organic or inorganic deposit, encrustations on ship hulls and natural surfaces in marine environment) (Nicholls and Hopkins 1993). Similarly, the bloom-forming algae *Gymnodimium,* which causes paralytic shellfish poisoning, was introduced into Australian waters (Hallegraeff 1992) from Japan. Extensive studies on the introduction of invasive species and their impact on native species exist in literature. Such studies on Indian ports and harbours are still in infancy and possibly, this is the reason why no reports of bioinvasion are available from Indian waters. However, recently there have been reports of the new record of a fusilier species *Casio cuning* from the coastal waters of Goa (Padate et al. 2010).

The occurrence of this species beyond its territorial jurisdiction could be referred to as ‘range extension’ wherein, the possibility of translocation of this species from elsewhere could be suspected.

Quantum of Traffic: The Mormugão harbour being a major port in India, is frequented by large tankers and cargo vessels in addition to international cruise liners and coastal traffic. The yearly traffic of this port on an average, includes about 60–125 coastal vessels, 30–40 international passenger cruise liners with capacities of 4,500–5,500 tourists per cruise, in addition to 400–500 cargo vessels (e.g. 521 in 2018–19, 430 in 2019–20) with capacities to carry up to 0.4 Mt dwt (deadweight tonnage is the measure of a vessel’s weight carrying capacity, which is the sum of cargo, fuel, fresh water, ballast water, provisions, passengers and crew, excluding the weight of empty ship) and/or bulk carriers with capacities varying from 1,200–52,000 Mm^3. The ballast tank capacity of tankers and bulk carriers exceeds 5000 m^3 though for passenger and general cargo ships, it may vary from 1500–5000 m^3. This provides a measure of the large amount of ballast water released in the vicinity of the Mormugão port.

Monitoring: The impact of discharge of ballast water from these vessels need to be monitored on a regular basis so as to minimize pollution of coastal waters and keep it under control to the extent possible. Since, once established, it is extremely difficult to get rid of an invasive species in the marine environment. It is important, therefore, to prevent the invasive species from entering the environment in the first place. As a preventive measure, one of the widely employed techniques to minimize bioinvasion is to discharge the ballast waters in open seas, what is generally known as ‘open sea exchange’.

Remedial Measures: In view of the rising concerns regarding the environmental impact of ballast water discharge, the International Maritime Organization in 2004 accepted for implementation the ‘International Convention for the Control and Management of Ships’ Ballast Water and Sediments’, to check the environmental menace from ballast water (BMW Convention 2004). According to this convention, ships are required to implement a Ballast Water Management Plan, which shall include the maintenance of a ballast water documentation book and implement the water management actions as per the outlined standard.

Threat from Microplastics: As mentioned earlier (refer Sect. 4.9.2) particulate plastics (<5 mm), the microplastics (MP) and the nanoplastics (NPt) are a major environmental threat to the oceans and aquatic life despite their tiny size. The poor management practices of solid waste and sewage from domestic and industrial effluents contain non-biodegradable waste such as plastics in addition to heavy metals, pesticides, fertilizers and others. As plastic waste proliferates around the world, it causes pollution by entering the natural ecosystem from a variety of sources including cosmetics, clothing, food packaging and industrial processes. It poses significant hazards towards microorganisms, humans, animals and plants. Among the marine animals, the shellfish is the cause of particular concern, as unlike fish, the entire animal, including the abdomen and the contained microplastics, is consumed. The microplastics have infiltrated all parts of the oceans and are now found in the smallest of planktons to the largest of whales. They have reached even the most remote regions of the planet, from Arctic sea ice to fishes in the deepest regions of the oceans, the Mariana Trench.

The microplastics form as a consequence of surface weathering and partial degradation by photolytic, mechanical and biological processes, of larger plastic materials such as plastic bottles, buckets, plastic toys, hair combs, toothbrushes among others, the so-called macroplastics, during which some of the leached chemical constituents get released into the environment. Likewise, the cosmetic and health and beauty products such as facial scrubs, cleansers, toothpastes among others contain the microbeads, which are very fine pieces of manufactured polyethylene plastic that are added as exfoliant to these health and beauty products. These get added through wastewaters and remain at large in the environment. The MP/NPt can also come from synthetic clothing such as nylon, spandex, acetate, polyester, acrylic, rayon among others. The threads of clothes that get carried away in the wastewaters during washing of clothes contribute to the addition of microplastics. Wastewater treatment removes most of the microplastics, yet a small percentage does get carried away into the water system.

Types of Microplastics: Two classes of microplastics are currently recognized (i) Primary microplastics and (ii) Secondary microplastics. The primary microplastics include any plastic fragment that is 5 mm or less in size such as microfibres, microbeads and nurdles that are manufactured for addition to certain category of products such as cosmetics, personal care products, detergents, paints among others. The secondary microplastics are formed from the degradation (breakdown) of larger plastic products through natural weathering processes-physical, chemical or biological, after their entry into the environment (Boucher and Friot 2017). The secondary plastics can also be produced during the production, transport and use of different plastic materials. The sources of these include water and soda bottles, fishing nets, tea bags, plastic bags, microwave containers, tyre-wear among others. In addition to these, secondary plastics can be generated by fragmentation of plastic waste from maritime, agricultural, industrial and urban activities.

These tiny particles easily pass through water filtration systems and get incorporated in soils and ultimately find their way to the ocean or are added directly to the ocean as waste. The MP/NPt due to their size prove much more challenging to eliminate from the environment compared to the macroplastics. The MPs can be captured only by using plankton nets which have mesh sizes measuring 0.1–0.5 mm, the NPts are still more difficult to isolate. It is estimated that in present times, there could be up to 5.25 trillion macro- and micro-plastic particles in the ocean weighing up to 2,69,000 t with fresh additions every day (4ocean Team 2020). According to the United Nations Environmental Assembly (2017), an estimated 4.8–12.7 Mt of plastics are introduced into the oceans annually. As per the current trends, the total plastic waste produced is predicted to rise by 33 Gt by 2050. About 35% of the ocean microplastics are derived from textiles and clothing, mainly due to the erosion of polyester-, acrylic-, nylon-based clothing often during the washing process. The microplastics also accumulate in the air and other terrestrial ecosystems.

Plastic Pellets: The raw material used in the manufacture of plastics consists of small granules, cylindrical or disk-shaped, with a diameter of a few (1–5) mm which are referred to as 'Plastic Resin Pellets' or as polypropylene (PP) resin pellets. The wide variety of plastic products produced internationally has created a demand for many different types of polymers or resins. An estimated 30 Mt of resins, most of which are formed into pellets are manufactured annually in the United States. India produced 6115 Mt of PP in 2019. These pellets are commonly called 'nurdles'. The most commonly used resins include polyethylene, polypropylene and polysterene. Recent studies have shown that the PP pellets were found to contain Polychlorinated biphenyls (PCBs), dichlorodiphenyledichloroethylene (DDE) and nonylphenol (NP). The PCBs are hydrophobic (water-repellent) substances able to accumulate in organic materials including plastic pellets. The DDEs are chemical compounds used to colour the plastic products. These compounds come in the form of dyes and pigments. The dyes are usually used with polycarbonates, polysterene and acrylic polymers, whereas pigments are better suited for use with polyolefins.

During the manufacture of plastic products, the plastic resin pellets are remelted and moulded into the desired shape to get the finished products. The pellets can get released to the environment both during manufacturing and transport. They are often ingested by marine organisms and birds. The adverse effects on organisms is a matter of great concern (Mato et al. 2001). The concentration of these compounds in PP resin pellets from oceans varied from 4–117 ng/g (PCBs) and 0.16–3.1 ng/g (DDEs) and 0.13–16 μg/g (NPs), similar concentration of these compounds were also observed in the suspended particles and bottom sediments collected from the same areas (*ibid.*). It was experimentally established that the source of PCBs and DDEs was ambient waters from which the compounds got adsorbed to the pellet surfaces.

Reservoirs and Toxicity: The high accumulation potential suggests that the plastic resin pellets serve both as a transport medium as well as a potential source of toxic chemicals in the marine environments. They consist of chemicals such as pesticides and dioxins which are hazardous to human and animal health in high concentrations. They are not very water soluble and hence it is easy for them to accumulate within aquatic sediments and create toxic reservoirs. The larger animals accumulate more of these toxins within their fat and tissue than smaller animals. This obviously leads to a more prominent risk factor for the large marine creatures and humans. Furthermore, the microplastics contain fillers, colourants, flame retardants and many other chemicals as a result of plastic processing. The rate of diffusion and dispersal of these different chemical compounds varies from compound to compound, meaning thereby, that some may pose less of risk than others.

Effect on Organisms: The microplastics serve as carriers for bacteria and persistent organic pollutants (POPs). They were considered as sources of E. coli and a pathogenic Vibrio bacteria which were noticed on five public bathing beaches in the U.S. as per a recent study (Mato et al. 2001). As the microplastics serve as bacterial vectors (Ma et al. 2020), non-native bacteria may enter fresh water or even marine ecosystems for instance from ballast waters, causing havoc on the pre-existing balance among living beings. Consumption of plastics is indeed harmful for marine animals, but ingesting bacteria-ridden material containing POPs is still more dangerous and could even be fatal. The microplastics affect not only the aquatic communities but they pose a range of issues to land-based ecosystems as well. They have been found inside the lungs of surgical patients and in human blood, though their tangible effects on humans are not yet known, however, they could be destructive to marine biota and logically humans being the part of the same global ecosystem may have to face similar adversities in future. The PCBs, DDEs and NPs are known to be carcinogenic and are banned by the Stockholm Convention (Mato et al. 2001). The U.S. Environmental Protection Agency too has classified DDE and others such as DDT, DDD as probable human carcinogens.

Rate of Degradation: As the rate of degradation of plastics is very slow, often over hundreds to thousands of years, the microplastics have a high probability of ingestion, incorporation and accumulation in the body tissue of many organisms. Thus, the toxic chemicals that arrive from the oceans and runoff could get biomagnified by movement up the food chain. In terrestrial ecosystems, microplastics have been shown to affect organism behaviour for instance, they reduce the viability of soil ecosystems and lessen the growth and therefore, the weight of the earthworms.

Methods of Quantification: The quantification of MP/NPt can be done by counting, weighing and from the absorbance and turbidity measurements of water. They can be analysed using scanning electron microscopy (SEM), Fourier transform, infrared spectroscopy (FTIR), Raman spectroscopy, surface-enhanced Raman spectroscopy and Raman tweezers (Hanif and Jalil 2022). The methods of mitigation differ depending on the removal characteristics: filtration and separation, electron nanofibre membrane; capture and surface attachment method: coagulation, flocculation and sedimentation, electrocoagulation, adsorption, magnetization, micromachines and microorganism aggregation and degradation: photocatalytic degradation, microorganism degradation and thermal degradation, where removal efficiency may vary from 58–100% (Hanif and Jalil 2022).

Mitigation: The best possible mitigation one can think of is to slow the spread of plastic debris at the source by reducing and wherever possible, avoiding plastic consumption/use. No major change in lifestyle is desirable, but small amendments and modifications in day-to-day behaviour can make a world of difference to the plastic menace. For instance, purchasing natural fibre-clothing, such as cotton and wool, employing the 'reuse option' for the reusable goods and recycling of plastics as per the guidelines of the local waste management, and using alternative transport, e.g. bicycles, wherever feasible, since vehicle tyres contribute microplastic dust to the atmosphere.

In addition, steps may be initiated to significantly curtail single-use plastic items, such as water bottles, plastic shopping bags, straws and other products. The state agencies need to strengthen garbage collection and improve recycling rate, to prevent waste from leaking into the environment. The scientific community needs to devise ways to break down plastic to its basic units, which can be rebuilt into new plastic or other materials. Europe recycles about 30% of the plastics whereas in the United States, it is about 9%. The need to increase recycling capacity is more now since China which use to import nearly 45% of all plastic waste intended for recycling, has stopped these imports.

Major Polluters**:** It has been observed globally that geographically and demographically large countries tend to produce more plastic waste in general, for instance the U.S. tops the list followed by India and China. However, in terms of per capita waste, it is seen that the smaller countries such as Micronesia, Bermuda, Palau, Hong Kong and others generate more waste than the larger, more populous ones. Also the safety precautions taken in the management of the waste are more in the developed countries than in the developing ones, as a result, inefficiently managed plastic waste lends up in the oceans. It is estimated that approximately 4.8–12.7 Mt of plastic waste enters the oceans every year. A 2017 study noted that about 80% of the waste came from just five countries namely China, Thailand, Vietnam, Indonesia and the Philippines, a similar conclusion was arrived at in the 2021 study as well. It also need to be noted that the developed countries such as the U.S., Canada, the U. K. and some others export the collected plastic waste to the Asian countries where it is then recycled or disposed of. This provides a distorted picture of the waste actually generated by both, the sending country as well as the receiving country.

Clean Oceans Initiatives: The Scandinavian countries in general, and Norway in particular, has taken the initiative to clean up the oceans since two-thirds of its export earnings come from oceanic and coastal activities particularly shipping, fisheries and aquaculture industries. Norwegian companies and research institutes have taken the lead in developing solutions and technology that can help to combat plastic pollution. The company Clean Sea Solutions has developed the Clean Sea Robot. This is an autonomous, electric, aqua drone that sweeps plastic waste from the ocean surface with the help of sensors (artificial intelligence and remote sensing). The collected waste is stored on board and when the devise is full, it returns to a dedicated station to be emptied and recharged for the subsequent missions. Research is on to develop models that can clean the deeper water columns of the oceans also. Other initiatives include chemical recycling of mixed plastic waste and transforming it into raw material for the manufacture of new plastic products. Likewise, an impressive 95% of all plastic bottles in Norway are returned via a national bottle deposit scheme, so that more waste is collected and recycled and less will end up causing plastic pollution. Research is in advanced stages to develop biodegradable materials to replace plastic nets.

5.3 Impact on Groundwater Quality

Information on the quality of water resource is essential for the protection of life and property, and for the effective management of water resources, required for economic well-being of the society.

Background: Groundwater often contains dissolved minerals and gases largely acquired from the soil zone or from the aquifer that give it the tangy taste. But for these, the water would be tasteless or flat. The presence of dissolved constituents along with the pathogens determines the quality of water.

The most common dissolved substances in water are sodium-, potassium-, calcium-, magnesium-chlorides, -bicarbonates, -nitrates, -sulphates and -phosphates. These are usually referred to as common constituents. The proportion of these constituents determines the quality of water. If the quantity of dissolved minerals/salts exceeds about 1000 mg/l, then the water is considered unsuitable for potable purposes. Water with a few thousand mg/l of dissolved minerals or their salts is classed as saline water. That with high concentrations of calcium and magnesium is called 'hard water'. Hardness of water is expressed in terms of concentration of salts equivalent to the concentration of calcium carbonate. Water is considered as 'soft' if it contains up to 60 mg/l of hardness, 'moderately hard' from 61 to 120 mg/l, 'hard' between 121 and 180 mg/l, and 'very hard' if the concentration exceeds 180 mg/l. Concentration of nitrate in potable water should be less than 10 mg/l. Similarly excessive iron can also be present in acidic water; it can be reduced by treatment.

Trace concentrations of certain elements are, however, beneficial to some plants and animals. The presence of boron in small concentrations is beneficial to some species of plants, but is toxic to others when present in slightly higher concentrations. A test of the acidity of water can be made from its pH (measure of the hydrogen-ion concentration. The pH scale ranges from 0 to 14. A pH of 7 indicates neutral water; that >7 alkaline, and <7 acidic). Alkaline water forms scales whereas acidic water can corrode. According to the criteria laid down by the US Environmental Protection Agency, water for domestic consumption should have a pH between 5.5 and 9.0.

5.3.1 Quality of State Groundwater

Background: Groundwater quality in the State has become a matter of great concern largely due to the increased anthropogenic activity. The groundwater in shallow coastal aquifers in particular, is more susceptible to bacterial contamination often referred to as *faecal coliform*. The faecal coliform are facultatively anaerobic, rod-shaped, gram-negative, non-sporulating bacteria. The presence of these is indicative of the presence of other pathogenic organisms that may include bacteria, viruses or protozoa and many multicellular parasites. The permissible limit for coliform in potable water is zero. In addition to this, nitrates and sulphates from the unsystematic/haphazard disposal of untreated organic waste, particularly the faecal waste, also affect the quality of groundwater.

Cause of Pollution: The groundwater quality in the State in general, can be classified as 'good' except in the coastal belt, where it has been adversely affected primarily due to increased anthropogenic activity. In this densely populated area, not all of the hotels, restaurants and public eateries have adequate drainage facilities. The wastewaters, sewage (animal excreta) and garbage, except in a few cases, are buried preferably in pits without any pre-treatment. This has been going on for over the last four decades especially along the coastal stretches of South Goa where temporary-shacks and tourist-huts burry human excreta into the ground. The highly permeable nature of soil and the shallow watertable conditions have facilitated the movement of pollutants in the subsoil zones. The net result is, the groundwater in the coastal aquifers is contaminated due to increased concentration of COD, BOD, phosphates, nitrates, sulphates, microbial pathogens (*faecal coliform*) among others, derived from sewage and decomposition of the organic waste, primarily faecal remains.

The groundwaters from Calangute, North Goa district for example, contain coliform ranging from 2,000 to 20,000 MPN/l, the permissible limits being 100 MPN/l. The nitrates and the phosphates too are much higher than permissible limits. The same groundwater is used by the residents except perhaps for the drinking purposes. Disinfection of groundwater as a pre-treatment therefore, has become necessary to safeguard the health of the inhabitants. The groundwaters in the vicinity of the stream courses in the capital city Panaji are polluted due to dumping of domestic sewage (CGWB 2013), in addition to pollution from the ingress of saline waters.

Treatment Facilities: As of now sewage treatment plants (STP) are available only at Panaji, Margao and Vasco da Gama with a combined capacity to treat 78.35 Ml/day. The total sewage water generated in the State is estimated to be 165 Ml/day (Anon 2021a). As regards the groundwater, an estimated 68.02 Mm^3 of the resource over an area of 430 sq km has been found to be of poor quality. It has been observed that even where sewage treatment plants are set up, as per statutory requirements such as in starred hotels, they are not operated regularly to cut on costs and the waste waters and sewage is discharged into the sea.

Sewage Leakage: Other pollution problems in the State include sewage leakage, faulty septic-tank operations and land-fill leachates. In coastal areas, intensive pumping of groundwater has led to saltwater intrusion in freshwater aquifers. Most towns are facing the pressure of groundwater contamination. The infrastructure to collect and treat sewage is inadequate and needs urgent attention.

5.3.2 Contamination of Groundwaters

Influence of Trace Metals on the Water Quality: Water from wells and springs in the vicinity of industrial estates contains a high concentration of dissolved minerals that are beyond the tolerance limits of humans and other animals or plants. As against, some spring waters, e.g. the one at Kesarval has trace concentrations of arsenic (?) and therefore, has medicinal property/value for the cure of some skin diseases. Contrary to this, the groundwaters from some wells tapping the granitic gneisses from Canacona reportedly contain more than permissible levels of heavy metals (?). The incidences of renal disorders (kidney disease leading to failure) have been noticed in the local populace for several years. On an average there are about 10–15 new cases of renal disorder every year, from this *taluka*. There is a possibility, however, that the kidney disorder may be related to frequent consumption of bivalve clams some of which are quite common in these sandy estuarine environments. They usually contain high concentrations of heavy metals. The exact cause of the renal disorder has not yet been established.

Contamination from Industrial Effluents: Increased urbanization of the tablelands has contaminated groundwater by a variety of industrial effluents especially when the industries are located on the lateritic plateaux, the effluents are detected in the springs and wells at the base of the plateau on the coastal plain (refer Sect. 6.3). There have also been instances of groundwater contamination by petroleum products (*Herald*, dtd. 05.03.2008) in the wells at Bogmalo and Sancoale (Mormugão *taluka*) villages located at the foot of the Dabolim *plateau*, where major storage facilities of aviation turbine fuel and of other petroleum products exist.

Toxicity of Groundwaters: In recent years, the growth of industry, technology, demography and excessive water consumption primarily for industrial applications, has increased the stress on both, the water and land resources. In a number of cases, the stipulated norms of effluent treatment and handling are violated. Thus, the quality of groundwater has been degraded due to the incorporation of toxic metals such as lead, arsenic, cadmium, chromium and others from the industrial effluents into the groundwaters. Municipal wastes, chemical fertilizers and pesticides not properly contained, have entered the soil, infiltrated some aquifers and have locally degraded the groundwater quality.

Groundwater Quality in the Industrial Estates: The groundwater quality of the wells from some of the industrial estates is monitored by the GSPCB (Table 5.1). Broadly quite a few physicochemical parameters are within permissible limits as per the Indian Standards of Drinking Water and Bureau of Indian Standards, yet the waters are not of potable quality, primarily due to the presence of coliform (Anon 2014).

Heavy Metal Contamination: Heavy metals such as Cr, Ni, Cu, Zn, Cd, Pb, Hg among others are well-known environmental pollutants because of their toxicity, persistence in the environment and the bio-accumulative nature. The processing of

Table 5.1 Groundwater quality data (yearly average) for wells in the vicinity of the Industrial Estatess (Anon 2014)

Location/Parameter*	Corlim	Kundaim	Pilerne	Sancoale	Verna	Zuarinagar	Cuncolim
Temperature °C	32.0	30.0	33.0	33.0	33.0	33.0	6.7
pH	5.6	5.9	6.1	6.8	7.1	6.7	3.6
Conductivity $\mu s/cm^2$	40.2	238.8	266.2	248.0	222.0	257.4	288.4
Nitrate mg/l	0.9	0.2	0.4	BDL	BDL	0.14	BDL
Dissolved oxygen mg/l	6.9	5.0	2.6	3.4	3.4	2.5	5.4
BOD mg/l	0.4	0.6	0.7	1.0	1.0	0.9	0.5
COD mg/l	BDL	BDL	BDL	BDL	BDL	BDL	4.0
Faecal coliform MPN/100 ml	–	–	–	–	02-540$	–	–
Total coliform MPN/100 ml	560	<1.8	<1.8	<1.8	33-540$	<1.8	<1.8

*All values are for the month of April, 2013; BDL: below detection limit $: Av. Nov.2013-Jan.2014

minerals/ores and their subsequent applications for industrial and economic purposes lead to the mobilization of these elements in the environment especially in the biological cycle. Accumulation of these potentially toxic heavy metals in the biota and in the aquatic and terrestrial food webs causes a potential health threat to their consumers including humans. It is therefore, necessary to monitor the concentration of heavy metals in the different segments of the environment.

The heavy metal pollution in the groundwaters from the Cuncolim industrial estate is a cause of concern (refer Sect. 6.3). In the majority of samples of groundwater from dug wells and bore wells, in and around the industrial estate, analysed for heavy metals showed Ni, Fe, Mn, Pb and Cd beyond permissible limits of drinking water (Anon 2014). The concentration of these metals need to be monitored on a regular basis and appropriate safeguards need to be adopted as may be required. Similarly, in the Bethora Industrial Estate hexavalent chromium concentration in excess of 10 ppm was detected in irrigation channel waters. However, no such detections were observed in the soil and groundwater samples (Anon 2016).

5.3.3 Saltwater Intrusion: Need for Efficient Management

With the increase in tourism activity during the last four decades, the groundwater resource in coastal areas is under extreme stress, and is vulnerable to saltwater ingress. The occurrence of saltwater intrusion has been detected in some borewells from several *talukas*, e.g. Pernem, Bardez from North Goa and Mormugão, Salcete and Canacona, from South Goa.

A number of dug- and borewell waters in and around Keri, Parascondem in Pernem *taluka*, Tivim, Colvale (Karaswada), Anjuna, Baga, Reis Magos and the stretch in the vicinity of Chapora and Baga river (Bardez *taluka*) is brackish due to the ingress of saline water in freshwater aquifer, as observed from the sampling of wells around Baga (Fig. 5.5) and other localities. Brackish to saline waters are also encountered in wells around Mormugão, those along the banks of river Sal (beyond the areas of normal tidal influence), at Velim, Varca and São José de Areal (Salcete *taluka*) and some wells at Palolem (Canacona *taluka*) in South Goa. The salinity levels are higher during the dry season, particularly in the month of May when there is minimal flow of fresh water and maximum ingress of saline water.

Although there is overexploitation of groundwater in the coastal areas and as such the shallow aquifers in particular, are vulnerable to degradation, the situation is certainly of concern, although may not yet be alarming, as far as the quantum utilization as well as the degradation of the water quality is concerned. However, there appears to be a mismatch between the demand–supply figures and the actual consumption of water, particularly in the areas dominated by the hospitality sector and industrial estates. Barely 12–54% of the resource is put to effective use in North Goa with maximum utilization in Ponda *taluka* (58%) followed by the Tiswadi *taluka* (54%) where the capital city is located. The least utilization of the resource is in Pernem *taluka*. In most southern *talukas*, the resource utilization is less than 30%

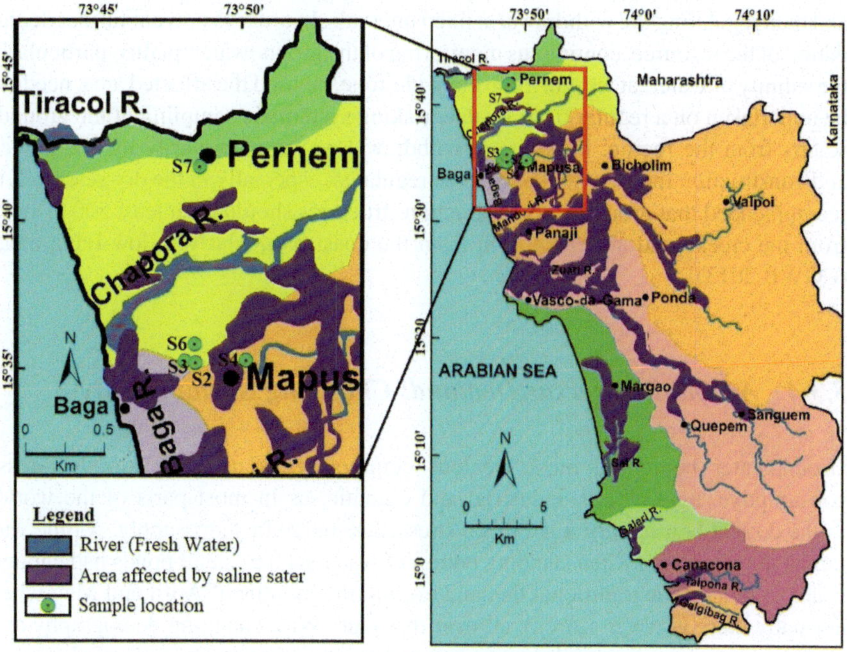

Fig. 5.5 Wells affected by saltwater intrusion at Baga, Bardez *taluka*

except in Salcete (42%). Thus, the resource potential is not fully utilized everywhere in the State and there is scope for development (CGWB 2019). As mentioned earlier caution need to be exercised in drawing such inferences, since the data available are inadequate and hence inconclusive for such generalizations. Considering the sustainable development of water resources, there is a need for systematic exploitation and efficient management of the groundwater resource. The development of the resource to cater to the overall water-supply needs should be carried out in an orderly manner based on laid down policies and strategies. Coupled with this is the need for awareness of the interdependence of surface and groundwater towards resolving water conservation and sustainability of the resource. This awareness will promote the establishment of an integrated water resource planning, development and management.

Mitigational Measures: As per the prevailing situation, the aquifers from the northern region of the State particularly the coastal belt of Pernem, Bardez, Mormugão and Salcete *talukas* have been experiencing saltwater intrusion within a fairly extensive zone of 0.29–0.30 km from the HTL (high tide line) and at places, such as São José de Areal (Salcete), the deeper confined aquifer is impacted. This situation is quite alarming and should be rectified in time before further damage occurs. Adequate remedial measures should be put in place to control and prevent further degradation of the resource, first by controlling excessive abstraction of groundwater and then regulating the withdrawal at fixed intervals. In order to prevent further degradation of the resource, continuous monitoring of the groundwater quality, particularly the salinity of water, at least in a one km wide zone, beyond the affected area, needs to be undertaken on a regular basis by undertaking systematic sampling of the groundwaters from the region. In addition, withdrawal of water primarily by the service and construction industry will have to be regulated, especially in the dry season. It is recommended that groundwater abstraction structures should be at least 500 m away from the creeks and at least 400 m from the coast particularly in low-lying areas (CGWB 2013).

5.3.4 Availability versus Demand: Changing Sustainability

Groundwater being the most precious commodity, is under quadruple stress-consumptive, agricultural, industrial and climatic, as in most parts of the world. In the coastal belt in particular, it is a stressed resource in short supply. A time was (about 40 years ago) when standing bodies of water such as small ponds had enough water to sustain the traditional demand even in the months of April and May when normally there is water scarcity all over the State. With changing demography and excessive pressures from a dominant hospitality, construction and industrial sectors, the surface water bodies get dried up even before the end of February. This is a good example of how sustainability of a resource changes with changing conditions, for instance in this case, the excessive demographic pressures. The balance of water demand is primarily met from groundwater.

As per the recent estimates by the Central Ground Water Board (CGWB 2019), net yearly groundwater availability is 160.33 Mm^3 and the total annual withdrawal is 53.71 Mm^3 whereas according to the Planning Commission, Government of India, the groundwater availability is 267.12 Mm^3 and the draft is 32.66 Mm^3 (Anon 2011). Although, there is disparity in these figures, they are for the State overall. Disproportionate withdrawal of water is, however, resorted to by the service sector (hotels, restaurants) and industrial units from the groundwater aquifers along the coastal belt as noted earlier. The stress on groundwater of selected areas of the coastal belt is also evident from the migratory pattern observed in the resident population of the State. The number of the urban centres has quadrupled to 44 since liberation (1961). The temporal demographic migrations largely due to relocation of rural population to the four urban metropolitan centres of Panaji, Margao, Mapusa and Vasco da Gama are presented below (Table 5.2) as an illustration. This trend has by and large continued although figures for the respective metropolitan centres for the decade in question are not available.

The water requirement of the four coastal *talukas* of the State, namely Bardez, Tiswadi, Salcete and Mormugão, (58% demography), is 220.19 Mm^3. This is partially met from domestic water supply 75.04 Mm^3 and the groundwater draft which contributes 30.15 Mm^3 and accounts for nearly 75% of the total groundwater withdrawal of the State.

Table 5.2 The temporal migratory pattern of demography to the urban centres of the State

Urban Centre	1961	1991	2001	2011	2021
Panaji	35,468	53,823	59,066	–	1,14,759
Margao	15,364	64,581	78,382	–	106,484
Mapusa	8,198	31,667	40,487	–	39,989
Vasco da Gama	6,483	83,367	97,154	–	100,000
Total	65,513	2,33,438	2,75,089	–	2,71,000

According to the Ministry of Irrigation 2000–2001 census, there are 4620 dug-wells, 60 deep tube wells and 30 shallow tube wells. Over the years the watertable has been receding at a significant rate, due to large quantum of withdrawal of water, particularly in the coastal belt (CGWB 2013). This is also noticed in the time series analysis of groundwater aquifers (ibid.) as mentioned earlier. Overall, it appears that the groundwater is underutilized and there is considerable surplus of water, mainly, since agriculture, the major consumer of groundwater, has nearly collapsed. The surplus may appear to be true on a regional scale and in the light of paucity of available data, as also apparent from Table 4.5. However, when seen in the context of saltwater intrusion and the quantum of water transported to some of these areas from elsewhere, one is constrained to think otherwise. The water transported from outside is not accounted for in the groundwater estimation charts, as discussed earlier. Therefore, the apparent surplus of water may be illusory. From the saltwater intrusion noted in several of these areas, which are the industrial hubs of the State, it becomes clear that excessive abstraction of water is indeed resorted to from localized structures of the industrial areas, for instance Karaswada in Bardez *taluka* and São José de Areal in Salcete *taluka* among others.

Adequate remedial measures should be put in place forthwith to control first, excessive abstraction of water and two, the resource should be monitored on a regular basis so that it is not tinkered with any further. In many cases, especially in the coastal hinterland saltwater ingress is noticed, on a limited scale, in restricted area of the stabilized dunes and the alluvial-colluvial deposits. The encroachment zone varies from about 250–350 m especially during low rainfall years, largely due to excessive withdrawal of groundwater. In addition, as mentioned above, extraction of sand from the estuaries also contributes to saline water encroachments in these unconfined aquifers. The requirement of water being large, the existing water bodies are unable to meet the demand. Water scarcity is especially observed during the dry season. High surface and sub-surface runoff, a consequence of rugged topography as well as the high permeability of the aquifers results in the lowering of water levels and drying of wells leading to water scarcity during the months of April/May. It is therefore, essential to monitor the potential saltwater encroachment zone, (about 2 km in width, from MSL) by determining salinity levels in well waters from the threatened areas at regular intervals. The dumping of domestic sewage (despite the existence of underground sewage) adjacent to stream courses in and around Panaji should be avoided, failing which, at least septic tanks which can retain the sludge should be made available, discharging the waters into the soak pits. This will help control pollution of groundwater to a large extent.

5.3.5 Quality of Surface Water

The two major rivers the Mandovi and Zuari which contribute significantly to the development of the State are naturally the ones that are the most polluted locally, followed by the Sal which is confined to the most populous, industrialized belt of the State. The National Green Tribunal (NGT) has used the Biological Oxygen Demand (BOD) values of <3 mg/l to identify the stretch of the river Mandovi between Marcel and Volvoi as a critical stretch. Eleven rivers/tributaries have been identified as being in need of intervention. The NGT has directed the State to prepare an action plan for the purpose of restoring polluted river stretches to the prescribed standards of water quality (Anon 2021b). Majority of the rivers in general, show high levels of coliform as they cross the industrial areas or the townships with total coliform levels varying between ~300–9900 MPN/100 ml (Anon 2019). The Sal river in particular appears to be the most polluted of all in the stretch between Panzorconi and Cavelossim (Salcete *taluka*).

Threat from Manganese to Potable Waters: The surface water bodies that cater to irrigation, industry and municipal water supply are all subjected to heavy contributions of particulate matter, especially during the rainy season, from the watersheds which also host mining activity. The Salaulim reservoir, which caters to the drinking water requirements of South Goa (water treatment capacity: 0.16 Mm^3/day), is drained by the Guloli river, a tributary of Zuari, that contributes manganese- and iron-charged waters to the reservoir. The levels of manganese in untreated water range from 0.21–0.28 mg/l (av. 0.26 mg/l), whereas in treated water they range from 0.05–0.10 mg/l (av. 0.08 mg/l) (Anon 2014). The desirable limit for drinking water as per the Indian Standard of Drinking Water (ISDW) specification- IS 10500:2004 specifications, is 0.1 mg/l and the permissible limit is 0.3 mg/l. The values of Chemical Oxygen Demand (COD) during the year range from 2–13 mg/l, BOD from 0.2–2.7 mg/l, dissolved oxygen from 6.4–7.6 mg/l are within permissible limits. The total coliform in the reservoir waters varies from <4.5–240.0 MPN/100 ml and faecal coliform from 0–170 MPN/100 ml (Anon 2014), however, the values (total coliform 490–4900 MPN/100 ml) increase considerably in the canal waters at Cuncolim. As for the Anjunem reservoir waters, the total coliform ranges from 79–920 MPN/100 ml, faecal coliform 23–540 MPN/100 ml. The permissible limit for coliform is 100 MPN/l. These values need to be controlled and drinking water supplies be made free from presence of coliform. The ISDW specification IS 10500: 2012 stipulates that drinking water shall be free from microorganisms such as algae, zooplanktons, flagellates, parasites and toxin-producing organisms.

Contamination from High Levels of Coliform: As far as the North Goa rivers are concerned, the Assonora river (supplies water to the Assonora water treatment plant, capacity 0.05 Mm^3/day) waters display an annual range of COD values from

1-11 mg/l, BOD from 0–2.8 mg/l and dissolved oxygen from 6.4–8.3 mg/l. Total coliform varies from 3,300–24,000 MPN/100 ml, whereas the faecal coliform shows a range from 680–13,000 MPN/100 ml. In the case of River Khandepar (e.g. Adyalkar 1985) at Opa (supplies water to the Opa water treatment plant, capacity: O.17 Mm^3/day), the COD values range from 1–10 mg/l, BOD from 0.2–2.4 mg/l, dissolved oxygen from 5.6–7.9 mg/l. Total coliform varies from 230–3300 MPN/100 ml and faecal coliform from 78–1,700 MPN/100 ml during the course of the year (Anon 2019). Such low values of COD, BOD and dissolved oxygen within permissible limits indicate fairly good quality river waters except that pathogen content is a matter of concern. The high values of coliform possibly suggest sewage contamination. The faecal coliform, in particular, are facultatively anaerobic, non-sporulating bacteria, the presence of which is indicative of existence of other pathogenic organisms that may include other types of bacteria, viruses or protozoa and many multicellular parasites. Bacterial contamination seems to be a common problem in most waters, which should be attended to by sewage treatment and adequate and appropriate disposal of sewage and human excreta.

5.3.6 Degradation of Coastal Aquifers

Traditionally, throughout Goa, in general (except in major towns) and in the coastal belts in particular, in the absence of any underground sewerage system or identified sites and agencies (except in recent years) specified for the sewage and garbage management, disposal of garbage used to be unsystematic and unregulated. This situation existed for several decades when the population density was low. The total population of the State then was less than one million. Over the years, there has been a progressive and sharp increase in demography due to migration which has left the traditional sewage and garbage disposal system in disarray.

Influence of Sewage on Water Quality: Sewage largely consists of 'grey waters' from sinks, baths, laundry among others and 'black waters' (toilet waters with human waste). In addition, open defecation also contributes to sewage-related pollution and is a serious concern of public health. Recently, 'Clean India' Mission /Programme (Swachh Bharat Abhiyan) has been launched nationwide with the intention of keeping the natural resource systems namely air, water and land free from pollution by faecal waste, muck and filth. Under this programme, the State provides sanitary toilets to households, particularly of the weaker sections. However, providing toilets is only a part of the problem of waste management. The more important aspect of management is proper waste disposal. Traditionally, untreated sewage gets combined with surface runoff and comes back and circulates in the environment, as sewers discharge into some standing body of water, such as the lake, river, stream, bay among others. So the breakdown of human waste is left to the ecosystem. In the coastal belt, particularly during the tourist season the human excreta is buried in pit-toilets (pit-latrines). The faecal waste percolates in the shallow, sandy aquifers polluting the groundwaters, in addition to the contamination from the domestic sewage.

In most modern towns and cities, however, the sewage is first treated in a sewage treatment plant (STP) and the treated waters are discharged into a body of water, thereby reducing environmental degradation. More efficient ways of grey water treatment can make the effluents not only usable but even potable as is being practised in Singapore (Anon 2021a). However, except for a few municipalities, most others from the State and the country as well, can neither provide underground sewage systems nor are equipped with STPs as they are expensive, unaffordable and given the financial status of most municipalities, are unviable. Thus, over 80–90% of the sewage remains untreated and contaminates the aquifers. Proper collection, adequate transport and nuisance-free disposal of fluid waste are a necessity of a developing community, to which the State is not fully equipped.

Wherever underground sewerage is unviable, septic tanks serve as an alternative. According to Census 2011, about 40% of urban households with access to sanitation are connected to septic tanks which retain the faecal sludge and the fluid is discharged through a soak pit. The sludge could be collected and sent for treatment to a STP. Thus provision for treatment plants should be the top priority of the State. Collection and transport of sludge could also generate employment.

5.3.7 Vulnerability Assessment

A wise and cautious groundwater management can slow down the degradation of the coastal aquifers, thereby preventing further damage and making them more resilient. The State should first recognize that the resource is threatened, and a susceptibility assessment should be undertaken lest it may be too late for redemption. The study will be able to identify aquifer susceptibility and the source of the contaminants. The assessment will enable the formulation of a sound policy for the development of the resource. Simultaneously, it will support better management of the programme and will provide an opportunity to formulate a strategy for exploitation. It will help put together the right land-use decisions in the coastal belt. More importantly, it will not only serve to educate the political establishment, but will also bring about an awareness of the impending threat to a vital natural resource. Needless to add, in all vulnerability studies a cautious approach should be adopted regarding the spatial and temporal variability and the processing of information. As is wellknown, in the absence of a reliable database, both a rosy as well as a bleak picture can be painted, as preferred. A careful approach will have to be taken therefore, to generate a realistic and sound database.

The assessment of the sustainability of the resource will help to categorize areas and devise strategies that would make it more resilient to endure the perturbation of contamination and overexploitation at the same time, enabling it to withstand the economic and ecological shock. A well thought-out approach on the part of the State would be crucial to avert the situation and lengthen the life of this precious resource. This can meaningfully add to a more sustained development of the region.

Water is a life-sustaining commodity and therefore is not taxed. However, for all purposes, other than sustenance, water needs to be treated as any other earth/mineral resource, the consumption of which should attract royalty to the State. In addition to revenue earning, it would also serve as a deterrent for overconsumption. This would serve to restrict the use of water leading to conservation of this precious resource. A small step in that direction has been taken by the State. A licence fee varying from ₹0.5–20.0 per m^3 of water is charged for industrial and commercial usage. Artificial reservoirs that cater to agriculture, industry and municipal water supply are all threatened by excessive mining activity in their catchment areas. In such instances, the principle of 'polluter pays' should be compulsorily made applicable to restrict the damage.

5.3.8 Mitigation Measures

The State should prevent some of the unhygienic practices and save the environment from further degradation and deterioration. As a remedial measure in most villages along the coastal belt, a sewerage system failing which at least septic tanks should be put in place. All hotels and restaurants, irrespective of the grade, should be compulsorily made to provide for septic tanks and soak pits for the dispersal of sewage and other effluents. Similarly, with the assistance of the State or local self-governing bodies such as the municipal councils or the 'village panchayats', the hotels (either singly or collectively) should be compelled to make arrangements for pre-treatment of the biodegradable garbage prior to the disposal. As a precautionary measure in the rapidly expanding hospitality industry, these statutory provisions should be enforced prior to the issuance of licence to the hotels and restaurants. There will thus be a check on the deterioration of groundwater quality.

The sludge from septic tanks being rich in nutrients, treated sludge could be used in agriculture as organic compost. Alternatively, it could be mixed with organic waste such as kitchen waste and used either for the generation of biogas or for manufacturing fluid/pellets and /or ethanol. Of late, a garbage processing plant (refuse-derived fuel process) with 100 tons capacity has been commissioned at Saligao and those of smaller capacity in other townships. These plants convert degradable waste to biogas (methane) which is utilized to generate electricity. The non-biodegradable, non-recyclable component is sent to the cement plants for co-incineration.

5.4 Air Pollution

Pollution of the atmosphere is a cause of concern since it can have adverse effect on human health, the ecosystem (environment), climate and precipitation (e.g. Akimoto 2003). Air pollution is caused by the introduction of inorganic, biological or any other fine solid particulate matter (referred to as PM), liquid droplets or gases into the air. These could be present individually or as a combination of particulate matter and gas in which case they are referred to as ***aerosols***. Burning of large quantities of coal and petroleum derivatives result in a mixture of smoke and sulphur dioxide producing smog which is a good example of air pollutant.

Type of Pollutants: Pollutants could be natural or man-made. They can be classified as '***Primary***' and '***Secondary***'. Primary pollutants are contributed by natural processes, such as volcanic eruptions which contribute large quantities of ash and gases to the atmosphere. Among the anthropogenic pollutants are included carbon monoxide from motor vehicle exhausts, sulphur dioxide, carbon monoxide, carbon dioxide and volatile organic compounds (VOCs) released from factories, thermal power plants and other industrial processes due to the burning of coal, gas and other multiple types of fuels. The VOCs are categorized as methane (CH4) and non-methane (NMVOCs) types. In addition, gases released from refrigerators, air conditioners, aerosol sprays among others that are referred to as 'chlorofluorocarbons' (CFCs) are also significant pollutants, harmful to the ozone layer and are currently banned from use.

Secondary pollutants are not emitted directly, but are formed in the air when primary pollutants interact with one another. Ground-level ozone is a good example of secondary pollutant. Anthropogenic aerosols currently account for approximately 10% of our atmosphere. In mining operations the major pollutants are dust (SPM: Suspended Particulate Matter), toxic metals and carbon monoxide. Among the SPM those that have the diameter of 10 μm (micrometres) or less are referred to as respirable suspended particulate matter (RSMP, includes dust), others include fine particles (diameter 2.5 μm or less, include emissions), ultrafine particles and soot.

In the mining belt the quantity of dust, referred to as suspended particulate matter, generated during mining operations and the transport of ore and overburden is tremendous. More than twenty thousand trucks are involved on a day-to-day basis in the ore handling and transportation activity within a restricted quadrangle measuring approximately 60 $\times$ 25 km. The vehicular traffic in the peak season is quite chaotic with overspeeding and overloading, with no concern for road safety. Stipulated traffic rules are quite often overlooked which is apparent from the number of road accidents that occur quite frequently (Fig. 5.6). The fugitive dust emissions generated by the friction of vehicle tyres with the surface of unmetalled roads, is tremendous. Each stage of ore handling, including excavation, loading, transport and unloading involves generation of fugitive dust.

The SPM concentration levels in almost all the operating mines exceed the prescribed limits. Dust fall rate in the mine areas and in the vicinity varies between 6 and 8 t/km^2/month and in many areas it exceeds the prescribed limit of 10 t/km^2/month. The RSPM in ambient air is reported to vary from 10–14% of the SPM and free silica content in respirable dust varies between 7 and 12% which is higher than the limit, namely 5%, specified in the guidelines of the Directorate General of Mine Safety.

State Air Quality: The air quality for various localities from the State is monitored by the Goa State Pollution Control Board (GSPCB). The annual average data (Table 5.3) for some selected stations from the mining belt and those from the coastal

Fig. 5.6 A bailey bridge collapse due to excessive overloading in the vicinity of Codli, North Goa

townships are presented below, merely as an example for comparison of the pollution levels before the mining ban and after (Anon 2010, 2014). If one considers an arcuate line passing through Assonora-Usgao-Ponda-Sanguem-Quepem, broadly to the east and south of this line the pollutants, particularly the levels of NO_2 and PM_{10} before the mining ban show double the level of concentration than those after the ban. In the townships too, the ratios of these two constituents are proportionately identical. At places, such as Bicholim, Codli, Honda, Usgao in particular, the levels of SPM exceed the permissible limit of 100 $\mu g/m^3$ during the period November to May when the mining activity used to be at its peak.

Impact of Dust on Plants: Deposition of dry dust on leaves of plants affects light absorption and consequently the photosynthesis. The attendant nutritional deficiency may not lead to death of plants, but may certainly affect the reproductive capacity. Settled dust also affects diffusion of gases into and out of leaves, as fine dust smothers (blocks the openings) the stomata. It also affects surface temperature of the leaves and in threatened areas, the entire community of plants is affected. Thus, the change in the structure and composition of the communities leads to change in the grazing pattern of the animals/livestock.

Dust Pollution from Coal Transport: The State imports coal to feed steel plants operating within Goa, Maharashtra and Karnataka. The coal is transported by rail, road and waterways. The rate of transport works out to about 25 t/minute. The coal unloading and loading at the harbour and subsequent transport to the end-user

Table 5.3 Comparative annual mean air quality data*for selected sites from Goa (after GSPCB; Anon 2010, 2014; averages are calculated for the period November to May of the respective year)

Location/ Constituent	NO_2 µg/m^3		SO_2 µg/m^3		PM_{10}µg/m^3		$PM_{2.5}$ µg/m^3	
	Before # (2010–11)	After #(2013–14)	Before (2010–11)	After (2013–14)	Before (2010–11)	After (2013–14)	Before (2010–11)	After (2013–14)
Panaji	19.4	7.9	6.0	5.0	102.2	43.5	na	na
Vasco	21.0	10.7	6.7	7.3	68.5	44.5	na	na
Margao	na	11.7	na	6.0	na	56.3	na	na
Ponda	na	0.5	na	5.7	na	49.9	na	na
Mapusa	na	12.3	na	10.1	na	58.1	na	na
Assonora	17.5	11.7	9.5	6.1	90.5	51.9	37.0	21.1
Amona	22.6	12.0	13.1	6.1	92.7	55.9	28.4	22.2
Bicholim	40.4	11.8	11.1	6.1	112.4	49.3	36.2	20.1
Codli	24.8	12.3	12.8	5.9	111.0	45.9	34.0	18.6
Curchorem	15.0	11.8	16.5	5.9	90.9	50.8	37.1	20.2
Honda	23.2	1.8	10.2	6.1	156.6	48.1	46.0	19.4
Usgao	15.0	11.5	5.8	5.9	223.4	51.5	77.6	20.3
Sanguem	na	11.6	Na	5.9	na	49.1	Na	19.5
Tilamol	na	11.6	Na	6.0	na	47.6	Na	18.4

* All figures rounded to first decimal place, na: not available # with reference to the mining ban

destination results in the generation of considerable air pollution at various stages, for instance (i) emission/release of particulate matter in the form of coal dust, both in air and water; (ii) emission of particles from the diesel locomotive exhausts; (iii) noise and vibrations by train movement; (iv) fires due to spontaneous combustion of coal, leave alone traffic congestion and road accidents (v) health impact of substances sprayed on coal to reduce dust emissions. The Mormugão port handled 24.7 Mt of coal in 2020. This figure is likely to increase to 51 Mt by 2030.

The Planning Commission, while formulating the tenth plan document of the nation, had recommended the setting up of a coal slurry pipeline to move large quantities of coal over long distances in the country. This mode of transport is both environment-friendly and also cost-effective in the long term. However, there are two limitations to the transport of slurry. One, it would require large quantities of water to generate the coal-slurry and two, the quantum of energy required to move the slurry over long distances. Finally, the release of used water from the slurry at the destination point and attendant pollution issues will have to be tackled. Instead, it was also suggested that electricity be made available to the unelectrified areas of the country.

5.4.1 Threat to Human Health

With the ever-changing requirements of the modern civilization, a geoscientist is today required to perform multifarious functions and practise his profession in wide-ranging sub-disciplines of the field of geoscience. One of the recent additions to these is concerned with the effects of geology on human health. This branch is included under the study of ***Medical Geology***. It is defined as the science dealing with the relationship between natural geological conditions and health factors in humans and animals, and understanding the influence of normal environmental conditions on the geographical distribution of such health problems (Selinus et al. 2005). It is mainly concerned with the adverse effects of the industrial development on the environment in general, and the health of the humans in particular.

Geologic factors influence the role of environmental issues that affect the health and well-being of the people the world over. Regrettably, there is a general lack of understanding on the relevance of these factors on animal and human health care among the general public, public health organizations and even the geoscience community. The Scandinavian countries have taken a lead in this direction and have raised an awareness of these issues worldwide. Some of the adverse effects of the inhalation of particulate earth materials on human health are as presented in Table 5.4.

Exposure to certain toxic trace metals could cause site-specific cancer as shown below (Table 5.5).

Mining is known as a killer industry due to the dangers to the health of the miners as well as other inhabitants of the area, primarily due to inhalation of suspended dust. Among the various sizes of dust, the most dangerous ones are those with less than

Table 5.4 Effects of inhalation of particulate earth materials on human health

Earth Material	Potential Source	Health Effects
Asbestos	Dust from various sources	Inhalation: Asbestosis
Silica dust	Mining and other industrial activity	Inhalation: Silicosis and several other diseases including lung cancer
Coal dust and coal fly ash	Coal mining, processing and combustion activities	Inhalation: 'Black lung' bronchitis, etc
Other mineral dust	Dust of talc, kaolinite, clays, mica	Inhalation: Mineral-specific fibrosis, such as talcosis, silicosis, asbestosis
Volcanic ash	Atmospheric particulates generated from eruptions	Inhalation: Irritation of the respiratory tract, asthma
Volcanic gases, fog, haze	SO_2, HF, HCl, other acid gases from active volcanoes	Inhalation: Irritation of eyes, throat, respiratory tract
Pathogens	Soils, and dust generated from soils that host pathogens such bacteria and fungi	Inhalation: Asthma and pathogens-specific diseases, such as valley fever. Percutaneous absorption: pathogen-related infections

Table 5.5 Effect of toxic metals on human organs

Element	Target organ and/or site
Cr	Lung, lever, kidney
As	Bladder, liver, lung, skin, vascular system, neurological changes
Cd	Lung, kidney, bladder, vascular system, hypertension, neurological changes
Hg	Nervous system, lung, kidney, brain
Pb	Nervous system, brain (IQ in children), anaemia

10 μm aerodynamic diameter (PM_{10}), since these pass directly into the blood stream on inhalation and get deposited on the inner wall of the respiratory system. Some of the adverse effects that have been noticed in the workforce that is constantly exposed to dust, are prevalence of silicosis, pneumoconiosis, bronchial asthma, emphysema among others. Incidences of chronic bronchitis and hypertension have been reported, however, this could be attributed also to socio-economic conditions and practices such as education, nutritional status, access to healthcare and so on. Sustained exposure to dust inhalation can cause pulmonary cardiac diseases, impaired foetal growth and infant development and many other health complications.

Similarly, the incidence of tuberculosis is more among the workforce from the mining areas, as supported by the number cases of patients referred to the Institute of Communicable Diseases, Panaji. The mortality rate per 0.1 M population ranged from 20.48–42.69 in the urban population and from 16.51–23.05 in the rural population between 1991 and 2001 (Kulkarni et al. 2007). In the South Goa mining belt, manganism has been noticed in several patients from the workforce. This disease is caused by intake of manganese-containing waters. It affects the brain and leads to loss of sanity. The symptoms are similar to those of Parkinson's disease. Similarly, due to high level of noise, 100–140 dB (decibel), prevalent in the mines, loss of hearing and deafness are common in the workforce.

As discussed in the preceding section, the impact of emissions on climate affects human health since it alters the composition of the atmosphere upon which life depends. Thus, humans are the drivers of climate change as well as the recipients of the impacts thereof. The capacity of societal development is considerably affected by changing pattern of precipitation, rising temperature, frequency of storms and similar other changes. However, the impact on societal development depends on the intensity of climate change and on the resilience or vulnerability of the society. Among direct impacts, probability of heart attacks, respiratory diseases, lung cancer and cardiovascular disorder is high. Indirect impacts, such as excessive rains and floods, wind, storms, drought, rising sea levels due to melting ice sheets may lead to loss of life and property and attendant spread of diseases such as typhoid, cholera, dysentery, diarrhoea among others.

5.5 Environmental Management

One of the prime requirements of environmental management is the continuous assessment of reserves, especially the recoverable reserves in relation to both the benefits likely to be accrued and the cost of damage on other natural resources such as forest, soil, agriculture, water, air and, humans and other organisms. The former is usually done by all mining concerns. What is more pertinent is the impact of recovery on the environment. In all studies pertaining to recovery, the most important aspect which is often ignored is the participation of the local community in the management of the resource. Unless this is attended to, any environmental management plan, however good, will be incomplete, since resource management is an activity that has a direct bearing on the local community, the stakeholder. This is essential, since the local traditional knowledge obtained from years of experimenting with nature goes a long way in formulating the planning and management process. In modern times, this could be achieved to some extent by undertaking a cost–benefit analysis of the mining activity as discussed later.

Mitigation Measures: It should be made mandatory on the mining concerns to take appropriate measures to rehabilitate the environment in the affected areas. This could include systematic and appropriate disposal of rejects so as to cause least impediment, appropriate stacking and preservation of soil that could be reused subsequently in already mined areas for afforestation/reforestation. This is essential, since reforestation of freshly mined material is not possible without adequate soil cover. New soil formation is a slow process. Hence to hasten reforestation, initially some soil will have to be made available to the mine rejects. Already mined areas could be refilled and covered by a layer of soil. This will help faster growth of vegetation. Normally, reforestation is avoided in view of additional costs. Re-vegetation of dumps then becomes a mere statutory formality.

The other aspect is stabilization of slopes which includes maintenance of appropriate slope angle. This is essential to prevent rock-falls and landslides. Mining normally leads to blockage of the drainage network. Alternate drainage channels should be made available to prevent flash flooding and inundation of surrounding inhabited and agricultural areas. Construction of retaining walls is a useful measure, which among other uses, also helps to channelize and control the movement of waters.

Mine-pit waters and beneficiation plant effluents must be adequately treated in terms of settling of suspended solids and clari-flocculation for reduction of inorganic load prior to the discharge. Maintenance workshops should be duly equipped for the collection of spilled oil and recycle it. Effluent treatment facilities must be augmented and existing ones improved to conform to prescribed standards for effluent treatment. Provision of safe potable water for the mine work-force and the inhabitants in the vicinity of the mine need to be made available. Establishment of a water treatment plant could substantially lessen the pollution caused by mine waters in the surrounding areas.

Aspect of dust pollution is often taken for granted. It is a matter of serious concern in mining areas, as long inhalation of dust could lead to serious pulmonary diseases. Dust should therefore, be controlled by repeated spraying of water which can lessen the concentration of suspended particulate matter in the area around the mines, areas along the road corridor and the affected localities in the vicinity of the mines where air quality limits exceed the prescribed norms (100 $\mu g/m^3$) for residential areas. The vehicles involved in transportation of ore should be duly covered. Avenue plantation should be encouraged. Green belts should be developed around mine pits and mine dumps. Wherever possible, alternate roads exclusively for ore transport should be constructed, bypassing the affected towns or villages.

Social and economic developments exert stress on the environment. There is a linkage between anthropogenic activity (economic activity), through stress effects (emissions, waste) to state of the resources (physical, chemical, biological) to impact on ecosystem, human health among others. The conceptual framework that depicts the relationship between the mining activity and the various components of the environment, is presented by means of a flow chart (Fig. 5.7).

5.5.1 Cost–Benefit Analysis

Any discourse that deals with exploitation of resources from the environment must include at least a guesstimate of the benefits accrued and the damages caused during the course of the activity, decision or programme. This can be achieved to some extent by undertaking the Cost–Benefit Analysis (CBA) which is one of the most accepted and widely used tools to quantify in monetary terms the value of the costs and benefits that are expected to result from the activity. The CBA enables to bring the various parameters/elements of an activity on a common footing or scale, so that

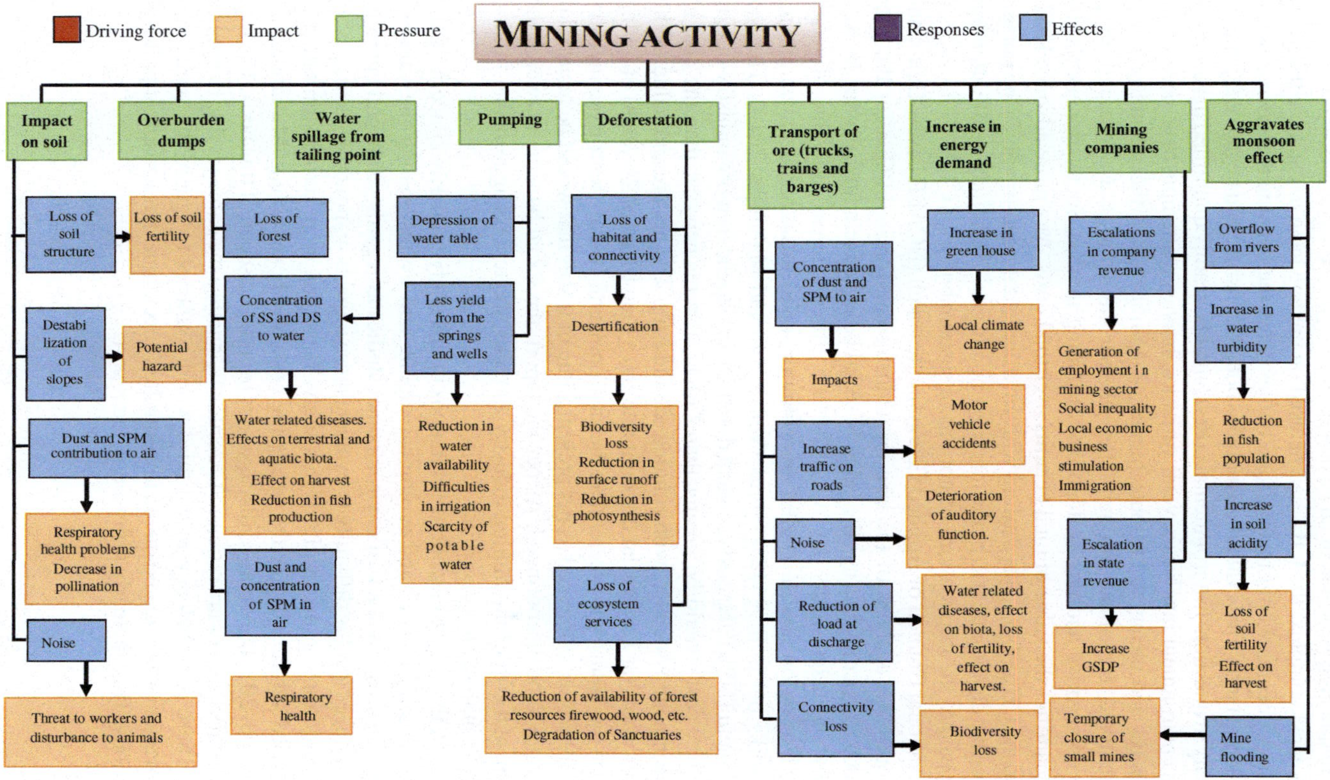

Fig. 5.7 Flow chart depicting the mining activity and its impact on the various components of the environment (modified after Pascual et al. 2013)

they can be compared with one another and thereby judge for oneself whether the activity in question was beneficial or otherwise.

At its most complex, the CBA can be used on the one hand to estimate the gains the mining concerns would receive, and on the other, the benefits that accrue to the community at large. This evaluation includes both the positive and negative aspects of the activity in question. Environmental costs cannot be readily quantified in monetary terms as environment involves several variables, however, attempts have been made to find ways and means to attribute monetary significance to the variables. It has been attempted to evaluate through CBA whether the net impact of the activity is positive or negative and this is done by calculating the Net Social Benefit (NSB) which quantifies the value of total benefits over total costs. At a more advance level, however, the CBA of projects dealing with environmental issues involves the pursuit of sustainable development to be the macroeconomic goal. Regrettably, little attention is being paid to the implications of notions of sustainability for CBA.

The National Council of Applied Economic Research (NCAER), New Delhi, has carried out CBA analysis of the iron ore industry and forests in Goa for the year 2008–2009. As per its findings the value of social benefits including revenue by way of royalty and taxes works out to be ₹2,309.5 crores whereas the value of social costs such as deforestation, landfill, air pollution, solid waste management among others works out to be ₹548.4 crores (Venkatesan and Majumdar 2010). The Net Social Benefit (NSB), the cost associated with giving up the iron ore industry in Goa (opportunity cost) would be greater than ₹1,842.2 crores per year than the cost to the environment associated with running the industry. The net present value of the opportunity cost (at 12% social discount rate and a time frame of 25 years) of giving up the iron ore industry is greater than the environmental cost of the iron ore industry by ₹14,449 crores, calculated as net present value of losses that would be incurred if mining were to be discontinued. As per the report, the social benefits outweigh the environmental cost of the mining industry and thus, the report suggests that mining should be continued in Goa (*ibid.*).

However, there are tremendous limitations to this study primarily concerning the underestimation of social and ecological capital, and overestimating the value of mineral resources. The lack of environmental studies as regards social and environmental damages and the difficulty to evaluate these impacts in monetary terms places serious limitations on the results of this study. The other, more serious limitation is the increased subjectivity for the intangible costs and benefits (e.g. Mukhopadyay and Kadekodi 2011). A detailed discussion is beyond the scope of this book. Clearly, further detailed studies are required to accurately analyse the damages due to mining in Goa. There is a tendency in the third world countries to ignore the social and environmental issues created due to uncontrolled development as a result of political compulsions. This is clearly reflected in the report.

5.5.2 *Rehabilitation of Mined Areas*

One often encounters that the scars of mining are visible for years to come much after the mining is over. This is a common sight in Goa and in other regions of the country where open cast mining is practised. It is only in such cases that scars of mining are glaringly visible. Mining by other methods such as underground mining often conceals the extent of damage. In the open cast mines the trenches and pits left after the extraction of ore need to be filled by mine rejects and then covered with a layer of soil thus making it suitable for the growth of vegetation. In the absence of soil, the mine rejects despite being made of fine earth, do not support the growth of plants.

Some of the abandoned mines and mine pits (Fig. 5.8) are put to alternative use such as pisciculture and recreational purposes. Exhausted mine pits could be utilized as rainwater harvesting structures. Some pits could be utilized to dump municipal and industrial waste after appropriate geological investigations of the terrain are carried out to ascertain the subsurface drainage, so that there is no adverse effect on the groundwater system and the local populace.

Recently Government of India has promulgated a 'Mine Closure Plan 2002' that emphasizes a planned rehabilitation programme which involves financial, sociological and environmental concerns of the mined areas. What is observed is, the abandoned mines are left in a state of decay and toxicity in a dilapidated condition. It has also been observed that mining companies abandon the mine not because the ore is exhausted, but because the cut-off grade/limit- the economic potential of the mineral deposit, is reached. It determines the cost involved in the extraction of ore. If the cost involved in extraction/transport and the like exceeds the market price of the

Fig. 5.8 Extensive, abandoned mine pit filled with water at Costi, South Goa

ore, then it may not be economical to continue mining under the prevailing market conditions. The mine is then subjected to alternate use with an impressive closure plan. Such initiatives need to be verified, monitored and prevented in advance by the concerned authorities so that precious resource is not lost forever.

5.5.3 Restoration Initiatives

In recent times, sporadic initiatives have been undertaken by some mining concerns from Goa towards restoration of mined areas. One of the ways is by reclamation of exhausted pits by backfilling and plantation on backfilled areas for soil-holding. Water-filled pit in which water remains throughout the year has been utilized for pisciculture on a pilot scale.

Plantations of certain species of plants and grasses on overburden dumps to control erosion and stabilization has been initiated. Invariably invasive plant species such as *Acacia auriculiformis* which has the best survival as well as growth rate, followed by *Casuarina alstonia* among others are planted. This gives rise to monoculture in the midst of dense tropical forest.

Construction of check dams for filtering of runoff mine waters has been attempted. Erection of bunds at the dump-toe to minimize wash-off is yet another initiative. Construction of lateritic bunds in a stepwise manner to arrest the movement of particulate matter such as silt and clay is also practised. Similarly, the dumps are covered by geotextiles to control slumping. Attempts have also been made to rejuvenate the abandoned fields in areas surrounding the mine involving the local community. Although these are small initiatives, they indicate the awareness of the mining concerns towards the restoration of the environmental damage.

5.5.4 Sustainable Mining and Protection of the Environment

Humans being the most evolved species on the planet, that have the ability to think and express, consider themselves at the centre of all activities and decide the destiny of all other living and non-living resources. They adopt and formulate policies that are beneficial to their existence and survival. The policy to mine the natural resources and formulation of the concept of environment are both creations of the humankind for its own benefits. It is inherent therefore, that imposition of ban on mining should not be seen as a solution, but merely a remedial measure to rectify the irregularities in the industry and streamline the mining operations. The very economy of the world depends on the mining of mineral resources as they contribute the raw material required to almost all types of industries. This means that humankind can seek development for its necessities of life and comfort solely through the sustained supply of mineral resources to the industry.

Protection of environment is also a prime concern of humankind. In fact, the concept of 'Environment' and the environmental policy is formulated by the world community solely for its own benefit. Safe and healthy environment is one of the prime requirements of civilization. All components of the environment such as air, water, flora, fauna and the earth systems as a whole are equally important to humans and in their preservations alone, lies their safety and well-being.

Thus, exploitation of natural resources and protection of environment are two facets of the same issue; neither of which can be neglected nor recklessly over-exploited. A balanced approach between the two is essential for the existence and survival of humans. All controversies, disputes, discussions and quarrels between nations, peoples and communities are attempts at striking the balance. There needs to be a conscious endeavour to use the natural resources in a planned and sustained manner. Improvement in technologies should be constantly evolved to enable the use of low-grade ores at low costs. Recycling of metals, use of scrap metals and other substitutes are some of the steps to conserve mineral resources for posterity.

Threat to Iron Ore Resource: Mining industry in Goa is largely concerned with exports. As the demand for ore increased in international markets and simultaneously emerging economies in the neighbourhood, such as China, picked up the developmental pace, the demand for ores increased manifold. Moreover, the Chinese markets preferred low-grade iron ores which were otherwise lying stacked as uneconomical. Multinational corporates diverted their attention to the mining industry in Goa. This led to a rapid pace of mining that exploited much more ore than the carrying capacity of the resource, concurrently some unscrupulous elements also played their role and large quantities of iron ore were illegally exploited from this tiny territory which has a limited ore reserve base and on which the economy of the State depended. Realizing the threat to the natural resource and the environment, the Hon. Supreme Court of India temporarily suspended all mining activity with effect from September 2012 and directed the State to formalize and regularize the mining operations in the State. In this process the question of protection of physical environment and ecology, ravaged by nearly six decades of incessant mining, must have weighed more over the socio-economic considerations, in arriving at a halt in mining. The mining ban was lifted in April 2014 with an assurance from the State government that adequate measures have been taken to resume mining operations.

During the ban-period, the State has witnessed considerable social unrest and economic strangle hold as large part of the population was dependent on mining activity. The government had to resort to severe imposition of taxes on all activities and commodities to counter the revenue loss from the mining ban. This led to additional stress on the local populace due to inflation. Surprisingly, the State remained completely silent on the issue of recovery of damages and costs from the mining concerns.

With the lifting of mining ban, the stakeholders were hopeful that situation would improve steadily to pre-ban days. However, this does not seem likely. The importing countries in the meantime have established contacts with new suppliers from other parts of the world. The price per ton of iron ore was slashed in international market from about US $150 per ton to about $ 45, largely due to competition and the global recession. However, the price (2022–23) has steadily increased to $102. Thus, presuming the mining activity and the export of ores do commence steadily, the ore is likely to fetch lower dividends than before the ban. It appears that the mining activity at best will be a little better than to break even, however, it may not be as profitable as before. An opportunity squandered is lost forever. It is therefore, essential that a measured, calculated and cautious approach is adopted by all concerned, so that situations of this nature are avoided in future and the country is not put to loss.

Sustainable Mining: The iron ore industry in the State is export-oriented and at present largely controlled by multinational corporates. The policy of corporate is to mine vigorously till the ore lasts and the mining activity is profitable. No sooner the company realizes from resource assessment studies that the cut-off limit is crossed, the mining activity is discontinued and the company closes down its operations in search of greener pastures elsewhere either within the country or abroad, as the case may be. This approach is beneficial to the company but not to the State. The State, on the contrary, should adopt a 'Sustainable Mining' approach especially considering the size of the resource, and potential benefits of mining activity, if it has to reap the benefits of mining in the interest of the State and local community. In comparison to iron ore reserves in other states of the country, the reserves in Goa are too small and in a way insignificant on the national scene, both in terms of quantity and quality for undertaking any extensive mining and even to establish a long-term export contracts with firms abroad. The importing countries too used the ores from Goa essentially for blending of ores imported from Brazil. Despite these limitations, the iron ores until the mining ban attracted the attention of purchasers from abroad, solely because (i) the iron ores in Goa enjoyed a 'place value' that is, proximity to a cheap means of transport, i.e. water transport, proximity to the port-head and nearness to an importing country, (ii) cost-wise the Goa ores were cheaper to the importing country compared to the South American ores and (iii) an established trade relationship already existed with the importing firms abroad, since pre-liberation times. This advantage that the iron ore industry enjoyed all these years has been now squandered and it will be extremely difficult for the industry to regain the lost confidence of importing firms from abroad in view of prevailing economic, geopolitical and environmental factors such as high carbon footprint.

5.5.5 Underground Mining and Excavations

Underground mining is a useful mining technique that can effectively hide the scars of mining, that are obnoxiously visible in the opencast process. Underground mining operations are by and large not prevalent in Goa as the chances of caving-in are very strong. On experimental basis, some adits were driven to win the ore especially where the ore body was located at deeper levels and the overburden required to be excavated was voluminous in comparison to that normally encountered. However, realizing the possibility of collapse (caving-in) the adits were abandoned in due course of time as a precautionary/preventive measure.

Underground excavations were, however, undertaken during the construction of Konkan railway tunnels primarily in the stretch from Ratnagiri to Mangalore. Several accidents due to collapses of tunnels have been reported with loss of human life. The tunnel excavations are in number of cases at least partly confined to the lateritized country rocks. In a typical laterite profile which may be tens of metres deep, a indurated laterite carapace called 'duricrust' or 'cuirasse' is underlain by soft clayey laterite which is often referred to as saprolite or lithomarge, this is followed by a weathered bed rock (protolith) which also has low competence. As the saprolite and even the weathered rock has a higher percentage of clay size material, both have high porosity and very low permeability. Often in rainy season the lithomarge is soaked with water and at times even a small vibration of the ground can trigger a landslide or a rock-fall as the water-soaked lithomarge is very plastic and thus can facilitate movement of the overlying hard carapace.

5.6 Construction and Industrial Mineral Resources

Construction and industrial minerals are the least known in the global mineral industry. They are essential for housing and infrastructure, yet they remain anonymous in the mineral industry, since their importance is governed solely by their 'place value' that is their location, proximity to a city, town or industrial establishment. If the extraction (the use of the term 'mining' is largely avoided here, since these activities are on a small scale and cannot be strictly included under 'mining', at best, they could be referred to as rat-hole mining) is undertaken far away from a town, city or industrial establishment, the augmentation in transportation cost is nearly 100% and the extraction of the commodity may become unprofitable. With the spurt in industrialization, the production of many industrial minerals is on the rise in developing economies and is an important source of employment. The attendant concern is the proximity of these excavations/quarries to populated areas which makes them a focus of land-use and environment with a direct impact on the cost of living.

Although almost all minerals are required by humans in some form or other, the few mineral materials discussed here include construction aggregates which consist of fragments of rocks and minerals that are used either alone or in combination with concrete, asphalt and plaster. These aggregates largely include sand and gravel from streams and rivers. Clays are also utilized on a small scale in the State by the brick and tile industry and at times even in cement industry. It is a potential material apart from porcelains and ceramics, in the manufacture of lightweight aggregate which finds applications in lightweight concrete sometimes required for the construction of buildings and bridges where weight is a matter of concern. Lime which has applications in several industries and which is made from material exploited from river beds is considered for scrutiny due to its wider environmental challenges.

In addition, crushed rocks particularly the mafic (dolerite/basalts) dykes intrusive in the country rock and located in the vicinity of towns and cities are subjected to exploitation for the purpose of road-metal aggregates. These materials are preferred in concrete due to their high compressive strength, however, with increase in demand for dolerite road-metal aggregate and the supply unable to keep pace with the demand, illegal/spurious trade is being resorted to. These deposit-locales are being squeezed out by other land uses presenting a problem to construction projects. Everyone wants cheap aggregates, but no one likes to have an aggregate mine/quarry in the vicinity of an establishment. The constraints on supply has resulted in the schistose rocks such as quartz-chlorite schists, and schistose metabasics, which are also dark grey in colour like the dolerite, are mixed with the dolerite aggregate. Although this may not cause much harm in road asphalts such mixed aggregates can have serious consequences in the construction of buildings, bridges and dams where weight and strength of the material is a matter of concern. Several intrusive rocks at St. Pedro at Ribandar in the vicinity of Panaji, and São José de Areal (Nessai) near Margao have been traditional suppliers of aggregate to construction sites, however, with the fast rate of urbanization most of these quarries appear to have a grim future.

5.6.1 Sand Extraction: Environmental Challenges

Degradation of rivers is the main cause of concern. In-stream mining lowers the stream bottom which leads to erosion and river bank collapse as seen in the case of Tiracol and Chapora rivers for instance. The depletion of sand in the streambed causes deepening of rivers and estuaries, and brings about changes in channel form such as widening of channel, and river mouths among others. The erosion of channel during high flow regimes of monsoon season is commonly observed. The degradation may lead to destruction of gravel bed, exposing the substrate that underlies the gravel, which could in turn affect the quality of aquatic habitat. The sand extraction can affect sediment mass balance in the river downstream, consequently inducing channel adjustment by incision, which may extend to considerable distance beyond the sand-extraction site. Such extractions have led to reduction in the size of the gravel and sand bars which is causing erosion of the downstream bars. Moreover, the mining of sand is a threat to engineering structures such as pier foundations of bridges and other structures. Hence, sand extraction/mining in principle is detrimental to the environment in general and hence is not recommended. However, if inevitable, mainly to protect the interests of the construction industry from the State, it may be undertaken with a note of cautious and on a limited scale after due geological and environmental impact assessment of the areas in question, is carried out.

Impact on the Stream Channel: The annual period of inundation, also called hydroperiod, largely controls the rate of sedimentation and degradation. The potential changes due to erosion, however, vary depending upon the geology and geomorphology of the terrain in question. The rate of erosion varies in different stretches of the river. In the higher reaches, especially in the mountainous/ghat sections, the erosional effects as a consequence of sand extraction downstream are not noteworthy. In the middle stretches of the river, should the sand extraction be excessive, than significant erosional effects may be noticed. This could be kept in control by keeping a check on the rate of extraction. This is particularly significant, as considerable proportion of sand derived from upland is trapped in the reservoirs constructed for irrigation and water supply. In the lower sections of the river (vicinity of the river mouth), the rate of erosion would be significantly affected, if the total quantum of sand extracted is more than about 25–30% of the total replenishment, despite the fact that a significant proportion of fine sand is contributed to the estuary from the ocean side due to the influence of longshore currents, particularly during storm surges.

Influence on Groundwater: Excessive extraction of sand, if resorted to, can have serious influence on the groundwater regime in areas in the vicinity of the streambed. In the estuarine stretches, the sediments have high porosity and permeability, as such the unconfined sandy aquifers from the coastal areas may be severely impacted, if the quantum of sand extracted is more than about 50% of the replenishment rate. Lowering of watertable may lead to drying of surrounding wells. Similarly, saltwater intrusion and other changes may be noticed in the vicinity. In particular, the coastal and near-shore environments are the most vulnerable. The effects of sand extraction in these transitional environments are compounded in recent times by the effect of sea-level rise.

Impact on Aquatic Biota: The physical impacts of sand extraction causes degradation of riparian and aquatic biota. As the sand extraction activity is largely confined to the estuaries, which are considered to be the most productive natural habitats, there is an implied threat to the aquatic ecology. It has the capability of causing major habitat disruption and may bring about a decline in biological diversity and productivity. A case in point is the decline in abundance of *Meretrix casta* (bivalve clams) from some estuaries, mentioned earlier. Enhanced turbidity levels may have an adverse effect on the biota and on the growth of mangroves.

5.6.2 Policy Framework

The State should put in place a policy framework for the extraction/mining of minor (industrial) minerals including sand, road metal, clays and other materials. This would enable concerned authorities to make statutory provisions for the extraction of these minerals. It would curb illegal extraction and would also facilitate extraction, monitoring and management of the resource as well as control degradation of the environment. The policy document *interalia* should ensure that (i) minor mineral extraction is carried out in a sustainable manner after careful study of the area, (ii) help to maintain river equilibrium by identifying locations and quantum of extractable sand/gravel/clay and (iii) assist minimizing the impact of extraction on the environment.

5.6.3 Guidelines

The Geological Survey of India has framed model guidelines in respect of mining of riverbed materials. These will be useful to frame guidelines relevant to sand extraction in Goa. In addition, the following points may be worthy of note. Extraction of sand may be permitted first from river sections which experience aggradation. The convex portions (inner side) of river bends, particularly upstream of estuarine sections, could be potential locales. However, the stabilized portions of gravel and sand bars (invariably vegetated) from palaeochannel sections may be strictly avoided. Similarly, downstream portions of sand bars, at river bends, could also be potential sites for extraction. The distance between sites for extraction shall depend on the sand budget. Sand and gravel may be extracted during dry season (e.g. October to May). The quantum of sand to be extracted shall depend on the width of the river and the replenishment rate. Estuaries being sensitive ecosystems, they should be avoided for sand extraction to the extent possible. If unavoidable, the sand and gravel extraction may be avoided at least from sections with potential for degradation/erosion. Extraction could be avoided at least within 1 km from crucial structures such as bridges, buildings, pumping stations and such other structures. Wherever the river cuts through its palaeochannels, the concave portions of bends contain gravel and sand. Extraction need to be avoided from such concave (outer side) portions. Manual extraction of sand alone may be permitted. Under special circumstances after due investigation of the local geological conditions, floodplain or off-channel extraction could be considered, making allowance for the rivers to replenish the quantity of material extracted from in-stream mining.

Requirement of Base Data: Potential river basins from the State have to be geologically investigated to identify prospective stretches of rivers and their tributaries, suitable for sand extraction. Information on topographic, hydrologic and hydraulic characteristics of the specific sites would be desirable. This information is required to determine the quantum of sand that can be exploited from the area without causing undue erosion and degradation, either to the site of extraction or areas upstream and/or downstream. In addition, base data will have to be generated on the river system such as areas of aggradation, potential sites of degradation, sand budget, thickness of the sediments within the channel/floodplain, nature and depth of the bedrock, nature, type and characteristics of the groundwater aquifer, flood discharge capacity of the rivers wherever there could be threats to infrastructure among others. An inventory of aquatic biota and riparian vegetation would be essential in framing statutes for sand and gravel extraction.

As of now, considering the need to maintain a sustained supply of raw materials to the industry and simultaneously minimize the impact on the river ecology, certain stretches (8 areas in North Goa and 4 in South Goa) of estuaries were identified by the State solely for sand extraction, without employing mechanized means of extraction. The depth of extraction and the site location within the channel were also specified. These areas need to be monitored by the State at regular intervals and illegal activity be made liable to penal action. However, these are provisional measures adopted in the absence of relevant geological information mentioned above.

5.6.4 Management Plan

A detailed management plan in respect of in-stream exploitation of shell resources is outlined under Sect. 5.7.6. The same is by and large relevant in the case of sand extraction with appropriate modifications. Additionally, following points may be significant. In-stream extraction/mining of sand should be primarily governed by the quantum of annual replenishment of the sediments from the provenance. The extraction should not exceed the sand replenishment. The maximum permissible depth of extraction should be less than the depth of the thalweg (the lowest point of the channel or valley) beyond which no extraction should be carried out. Extraction activity should be permitted to the extent possible, to downstream sand and gravel bars (not stabilized) at river bends. Cumulative effect of sand mining on the natural environment needs to be evaluated. The extraction activity should be restrictive so as to minimize the area of disturbance. There should be reduction in activities that release fine sediments to the river. A riparian buffer at the edge of water and against river bank should be maintained. The resource should be efficiently managed in a sustainable manner involving all stake holders.

5.7 Fossil Marine Shell Extraction: Environmental Concerns

'In-stream' fossil-shell extraction is being carried out manually by the local communities, from the estuaries of most of the rivers from the State, since several decades. In that sense it could be considered as a traditional activity. However, as the activity is largely confined to the estuaries, which are considered to be nurseries for fish breeding, there is an implied threat to the biota and the environment. As of now, due to eco-sensitive nature of the resource, exploitation is banned. The following section underscores the areas that are likely to violate applicable/existing environmental standards/parameters should the exploitation be permitted.

Shell extraction activity being solely confined to the intertidal zone, it includes large tracts of mangroves. Being a coastal stretch, activities in the area are governed by the Government of India notification of the Ministry of Environment and Forests issued under Section 3(1) and Section 3(2) (v) of the Environment Protection Act, 1986 and rule 5 (iii) (d) of environmental (protection) rules 1986. Any mining type of exploitation activity in this stretch is likely to have an impact on the environment and therefore attracts the provisions of the Act.

Some of the issues pertinent to these deposits are as follows: (i) the excavation may lead to pollution by contribution of particulate matter to the water system as suspended load. The concerns are (a) the loosening of the semi-compacted sediments may lead to erosion by traction and (b) contribute particulate matter as suspension to river waters. This could create turbidity in water thereby adversely affecting the biota and the loosened material could be washed away, i.e. eroded both by traction and suspension (ii) the physical alteration to the excavated site may lead to increase in water depth and stronger current flow that may increase erosion of land adjoining the estuary, (iii) the excavation may lead to destruction of mangroves and suspended load may have adverse effect on their growth and also on the growth of biota (this is particularly important since these coastal wetlands are considered to be nurseries for fish and shrimp culture), (iv) the stacking of overburden and the likely effects of its dispersal on the surrounding land and lastly (v) the excavation could lead to loss of geological record of this area.

5.7.1 Geological Perspective

The siltstone, that binds the shells together, consists of silt-size quartz whereas the clay fraction consists of minerals belonging to the mixed-layer clay group. The minerals of this group have the property to swell by adsorbing water and gain plasticity. The clays therefore, play an important role as binders to the silt. As such, when excavated the silt grains rarely have the opportunity to get separated from clay for ultimate removal either by traction or suspension in river waters. Moreover, as the excavation/extraction is confined to the inter-tidal region, water movement, both during high- and low-tide is gradual, as such water currents capable of eroding the clayey-silt are precluded. Erosion by traction is not envisaged as the silt is bound together by plastic clays. The swelling and plasticity will enable least contribution to the suspended load. Suspended load being minimal, adverse effects on biota are considerably reduced. Air pollution is not an issue as the excavation is carried out manually in water-wet condition.

Although loosening of material during extraction is likely to occur, this would not pose major erosional problems as the excavated material being water-wet settles under its own weight provided the rate of extraction is slow enough as in case of manual exploitation. The shells are separated from the silt by sifting in situ, within the pit/well at the time of extraction itself. Thus, the slurry generated by sifting is retained within the pit/well thereby precluding erosion and reducing contamination of the surroundings. Problems of stacking of overburden is not visualized as the void created by removal of overburden from one pit could be filled by the overburden from the adjoining pit/well and thus the problem of subsequent dispersal of particulate matter is negligible. All these provisions are valid subject only to manual extraction.

5.7.2 Habitat

It has been observed that the growth of mangroves is luxuriant in a narrow zone/belt parallel to the present day river (estuary) course; a few tens to hundreds of metres wide, along either banks. A diverse group of fauna, as is generally known, is also associated with the mangroves. Whereas beyond this zone, the region is cleared of mangroves due to anthropogenic activity primarily agriculture and pisciculture (in recent times shrimp farming) or rarely the mangroves are extremely sparse. Thus, vegetation is minimal in areas of potential excavation sites. Mangrove-vegetated areas at most could be delineated and declared as 'protected areas' instead of declaring the entire region as a protected area (e.g. Notification of the Ministry of Environment and Forests, Environment Protection Act 1986). Exploitation could then be solely confined to relatively barren areas. Violation of this could be checked by periodic inspections. However, this does not seem to be a permanent solution since the existence of mangroves and their proliferation depend on the stability of the ecosystem of the entire region.

5.7.3 Employment Opportunities

The traditional exploitation activity provides employment to the local population, who are the local stakeholders, thereby improving their economic status. However, due to the ban on exploitation, they are without gainful employment and thus there is considerable discontent. The shell deposits are located within the inter-tidal region where they are constantly subjected to the action of salt water. This leads to leaching of Ca and thus has an adverse effect on the quality of the deposit leading to its degradation. If unutilized, the resource will be rendered worthless within foreseeable future.

Traditional local exploitation in a way aids in the conservation (sensu Mineral Conservation Policy 1993) of this natural resource. Natural resource conservation, as it is understood today, is not mere protection and preservation for posterity but it involves optimal utilization in such a way, that the use of the resource for the production of goods and services does not unduly disturb its potential to meet the needs and aspirations of the future generations. Conservation management should ensure sustained productivity for meeting the present and future needs of the community that depends upon the resource. This naturally implies that if the present resource is unutilized in the foreseeable future, it is likely to be lost forever. This is not in keeping with the mineral conservation policy. On the contrary, ensuring sustainable utilization will provide employment and livelihood to the local people.

5.7.4 Statutory Framework

The shell deposits of Chopdem-Tuem, North Goa are located within the inter-tidal zone and are protected by the notification of the Ministry of Environment and Forests issued under Section 3(1) and Section 3(2)(v) of the Environment Protection Act, 1986 and rule 5 (iii) (d) of environmental (protection) rules 1986. The exploitation of these deposits is covered under Note-2 of Prohibited Activities: (ix) which states 'mining of sands, rocks, and other substrata materials, except those rare minerals not available outside the CRZ (Coastal Regulatory Zone) areas'. Moreover, mangroves are categorized under CRZ-I.

5.7.5 Environmental Strategies

Two types of strategies are normally followed to manage the resource and the environment. One is a 'top-down' policy implementation strategy. This is a more traditional strategy and also fairly effective to protect the environment. It is a useful approach that is very effective for a short period especially during early stages of implementation, but invariably fails in the long run, as is also seen in the present case. For instance, despite the ban, illegal exploitation of shells continues unabated. The other type of strategy involves the 'bottom-up' approach that involves participation of the stakeholders. This is a more practicable and effective approach that is usually followed currently in most cases of environment management. In this approach, once the stakeholders are convinced of the benefits of recommended management practices, there is voluntary participation of the stakeholder community which is the key to its success.

5.7.6 Environment Management Plan

Should the Government decide to bring in suitable ammendments to the notification of the Ministry of Environment and Forests, Environment Protection Act 1986, thereby making the deposits exploitable, following recommendations may be worth taking note of: (i) the ban imposed on exploitation has led to discontent among the local population. In due course of time, it could lead to local resistance and fuel militant attitude due to lack of employment opportunities leading to societal problems; this should be avoided, (ii) the shell deposit may be prospected by detailed mapping and surveying to evaluate the exploitable reserves. The carrying capacity of the resource need to be assessed and exploitation, being a traditional occupation, be permitted and progressively phased out. Advance action is necessary to identify alternative vocations or augmentation of other employment/vocations in the surrounding areas, (iii) manual extraction is recommended to minimize erosion, as the rate of excavation is slow and the volume of material excavated is limited. This will therefore, provide enough time for the water-wet sediments to settle and stabilize, thereby precluding silt erosion. Mechanized exploitation/mining is not recommended, as faster rate of excavation and larger volume of overburden would not allow enough time for the rejects to settle and stabilize. This would therefore, lead to larger proportion of particulate matter likely to be transported as traction-load and as suspension leading to water pollution, (iv) the licence/lease for exploitation may be issued initially for a period of not more than three years, solely for the exploitable reserves within the licence/lease area. Further, renewal for a period of one year at a time may be considered after inspection of the area by a competent authority and ascertaining satisfactorily that no environmental degradation has taken place, (v) periodic inspections of the mining lease areas should be undertaken to check for violations and infringement. Reconnaissance, surveillance and monitoring should be carried out since the estuary is an important ecosystem, (vi) enforcement of legislation should be adhered to scrupulously, (vii) the current trend in management of resources is to follow sustainable development involving all stakeholders. A combination of 'top-down' and 'bottom-up' (participation of stakeholders) approach may prove to be more successful. This approach may be adopted to ensure local community participation and to sustain their livelihood, (viii) delineation of mangroves on either side of the present-day river channel should be carried out before granting permission to issue exploitation licences. These zones then, may be declared as 'protected areas', (ix) an inventory of living resources and the uses that they are being put to, need to be undertaken to monitor the adverse effect of excavation, if any, (x) the conservation and consolidation of genetic resources of plants and animals should be undertaken, (xi) measures for sustainable harvesting of natural resources should be worked out, (xii) restoration and protection of natural vegetation and promotion of natural regeneration need to be attempted, (xiii) soil conservation work coupled with planting of mangrove vegetation should be undertaken in the inter-tidal region, (xiv) changes in ground and surface waters should be monitored with the help of remote sensing and GIS techniques, (xv) shifting of river course due to monsoon flooding should be controlled by proper embankment, this will help control erosion along the banks, (xvi) pollutants, if any, from the surrounding industries need to be monitored as it could have an adverse effect on the flora and fauna; estuaries being the nursery grounds for fish and shrimp communities, (xvii) certain area should be reserved as 'geological monument' that could serve as reference site for posterity, (xix) community-based activities (education, demonstration of alternative construction materials, participatory projects) should be encouraged to relieve pressure on this estuarine resources, (xx) the long-term solution to avoid exploitation of such estuarine earth resources and thereby safeguard the environment, is to promote alternative materials such as good quality limestone for the manufacture of lime (CaO) and offer alternative sources of livelihood to those employed in exploitation and (xxi) a periodic review of the impact on the environment may be undertaken, e.g. after every five years.

5.8 Concerns of Rising Sea Level

Tourism in Goa is primarily concentrated along the beaches which therefore, need to be protected and conserved from vagaries of changing climatic conditions. It is well-known that the coastal areas of India and most parts of the world are threatened by sea-level rise (National Research Council 1990) and Goa is no exception. Many attribute the rise to global warming, and there are strong arguments to the contrary as well. Rise in sea level is not an unusual phenomenon for the earth scientists. Both the rise and fall of sea level are long-term phenomena that operate on the scale of thousands of years, if not millions. Broadly, the Goa coast is one of submergence as evidenced by the Altinho hillock at Panaji that served as an island during historic times (Feio 1956). Evidence of subsequent recession of sea is noticed at several places including the beach rock at Polem in South Goa and at Vagator-Anjuna in North Goa. Rise and fall of sea level are very slow processes. The intervening time period is generally marked by small cyclic oscillations of the sea level which are locally observed. Thus, the possibility of a sudden sea-level rise or fall by a couple of metres or so within even a decade seems to be a remote possibility.

Therefore, the present rise of the sea level will be followed by a recession of the sea only to be followed by another incursion. Such oscillations that take place over intervals of decades ultimately leading to either an overall rise or fall in the level on the scale of millenia. Such instances have been many in the geological past. Nevertheless, the rise in the level observed at present need not be taken lightly or ignored in the interest of safety to life and property both of the local populace as well as tourists (Fig. 5.9). At the same time, it need not lead to pressing the panic button.

5.8.1 Vulnerability Assessment

On a national scale, of the total coastline of 7,500 km, about 3,000 km is reportedly been rendered rocky due to shifting of beach sands. Nearly 27 km long stretch of the 139 km coastline of the State has been subjected to erosion (*Times of India* dtd. 14. 04. 2022). Although largely attributed to rise in sea level due to global warming, some have ascribed it to construction of new ports and harbours along the coast. It is feared that the situation is likely to worsen if clearance is given to construct additional harbours of which about 187 are in the process of being sanctioned. It must be pointed out that the erosion observed along the coasts is not entirely due to the construction of new ports, although some contribution from it cannot be denied. It is more due to change in the currents pattern that is associated with the change in the climatic conditions and more so to the local rise in sea level attendant to changes of the continental shelf of the region in question. Nevertheless, appropriate measures could be set in place in the eventuality of a potential threat due to the sea-level rise. One of the measures is to identify potentially vulnerable areas. This should be done on the advice of a panel of experts drawn from the Earth

Fig. 5.9 Beach degradation due to rising sea level, Sinquerim, North Goa

science community who would assess the coastal region from the point of view of its geological and geomorphological makeup and formulate a strategic plan to take care of such eventualities.

References

Adyalkar PG (1985) Hydrology of Goa with particular reference to the lateritised midlands and the coastal strip. Proc. Seminar Earth Resources for Goa's Development, Geol Surv Ind:484–494

Akimoto H (2003) Global air quality and pollution. Science 302:1716–1723

Anon (2010) Annual Report for the year 2009–10. Goa State Pollution Control Boards, p 97

Anon (2011) Goa Development Report, Planning Commission, Government of India, New Delhi, Academic Foundation, New Delhi, pp 1–250

Anon (2014) Annual Report for the year 2013–14. Goa State Pollution Control Boards, p 256

Anon (2016) Pilot survey for disseminating SME's Technologies for groundwater remediation by bioremediation of the Sal river and Bethora Industrial Estate in Goa, pp 1–11. Doi:openjicareport.jica.go.jp

Anon (2019) Annual Report for the year 2019–20. Goa State Pollution Control Boards, 1–469

Anon (2021a) Official Gazette, Government of Goa, Series 1, No. 23, 1213–1244

Anon (2021b) Annual Report for the year 2019–20. Goa State Pollution Control Boards, pp 1–354

BMW Convention (2004) Implementing the Ballast Water Management Convention (in force from 2017). International Maritime Organization (IMO). Doi:imo.org

Boucher J, Friot D (2017) Primary microplastics in the oceans. Doi:https://doi.org/10.2305/IUCN.CH.2017.01.en

Brown BE (1997) Integrated Coastal Management, South Asia, Department of Marine Sciences and Coastal Management, University of New Castle, Hindson Print, Newcastle-upon-Tyne, UK, 1.1–10.17

Buis A (2019) The atmosphere: getting a handle on carbon dioxide. Doi: climate.nasa.gov

Central Ground Water Board (2013) Ground water information booklet, North Goa district, Goa State, Southwestern Region, Bengaluru, pp 1–22

Central Ground Water Board (2019) Aquifer mapping and management of groundwater resources, Goa state, Southwestern Region, Bengaluru, pp 1–24

Dessai AG (2018) Geology and Mineral Resources of Goa. New Delhi Publishers, New Delhi, p 323

Gordon JE, Baron HF (2013) The role of geodiversity in delivering ecosystem services and benefits in Scotland. Scott J Geol 49:41–58

Gray M (2005) Godiversity and geoconservation: what, why and how? The George Wright Forum 22:4–12

Feio M (1956) Problemas geomorphologicas de Goa. Garcia De Orta Rev Junta Misc Geog Investi Uram 7:1–39

Hallegraeff GM (1992) Harmful algal blooms in the Australian Region. Mar Pollut Bull 25:186–190

Hanif MA, Jalil AA (2022) Microplastics and nanoplastics: recent literature studies and patents on their removal from aqueous environment. Sci Total Environ:810. https://doi.org/10.1016/j.scitotenv.2021.152115

Kulkarni MS, Vaz FS, Motghare DD, Ferreira AM (2007) Trends in tuberculosis mortality rates in Goa (1991–2001). NTI Bull 43:9–13

Ma Y, Wang L, Wang T, Chen Q (2020) Microplastics as vectors of chemicals and microorganisms in the environment, In: Bolan NS, Kirkham MB, Halsband C, Nugegoda D (eds) Particulate plastics in terrestrial and aquatic environments. CRC Press, p 466. Doi:https://doi.org/10.1201/9781003053071-17

Mato Y, Isobe T, Takada H, KanehiroH OC, Kaminura T (2001) Plastic resin pellets as a transport medium for toxic chemicals in the marine environment. Environ Sci Technol 35:318–324

Minchin D, Cook EJ, Clark PF (2013) Alien species in British brackish and marine waters. Aquat Invasions 8:3–19

Montgomery CW (2006) Environmental geology. McGraw-Hill, New York, p 540

Mukhopadhyay P, Kadekodi GP (2011) Missing the woods for the ore: Goa's development myopia. Econ Pol Wkly 46:61–67

National Research Council, Division on Earth and Life Studies, Commission on Geosciences, Environment and Resources, Geophysics Study Committee (1990) Sea-level change. National Academies Press, p 256

Nicholls KH, Hopkins GJ (1993) Recent changes in Lake Erie (North Shore) Phytoplankton: cumulative effect of phosphorous loading reductions and the Zebra Mussle introduction. J Great Lakes Res 19:637–647. https://doi.org/10.1016/S0380-1330(93)71251-4

Padate VP, Rivonkar CU, Chandrashekar AA (2010) A note on the occurrence of reef inhabiting, red bellied yellow tail fusilier, *Casio cuning* from outside its known geographical array. Mar Biodivers Rec 3:1–16. https://doi.org/10.1017/S1755267210000151

Pascual XL, Cascallar MP, Planell i Calle M, Rodes EP, Carbonell CS, Vile i Casau A (2013) Iron mining in Goa (India) an interdisciplinary study, Master degree dissertation, Autonomous University of Barcelona, Spain (unpublished)

Seleinus O, Alloway B, Centeno JA, Finkelman RB, Fuge R, Lindh U, Smedley P (eds) (2005) Essentials of medical geology: impacts of natural environment on public health. Elsevier Academic Press, Burlington, p 812

Shikazono N (2012) Introduction to earth and planetary system science. Springer, Japan. http://link.springer.com

Tukiainen H (2019) Multiscale relationship between geodiversity and biodiversity across high altitude environments: implications for nature conservation. Nordia Geographical Publication, p 48. http://www.sciencedirect.com

UN General Assembly (2017) Focussing on People: Striving for Peace and a decent Life for All on a Sustainable Planet, Regular Session of the UN General Assembly 12th September, 2017, New York. http://sdg.iisd.org/events/72nd-session-of-the-un-general-assembly-unga-72/

Valdiya KS (1997) Geology, environment and society. Universities Press (India) Private Limited, Hyderabad, p 222

Venkatesan R, Majumdar R (2010) A study of contribution of Goan iron ore mining industry. National Council of Applied Economic Research, New Delhi, pp 1–70

4ocean Team (2020) What are microplastics? And why you should care. http://www.4ocean.com

Chapter 6
Waste Management

Abstract Management of the coastal zone of the State has been a serious concern both due to population pressures and preferred site for urbanization and settlement, consequently the land resources are under extreme stress. Increased solid waste, excessive pollution and augmented emissions are the other threats to the environment. Waste disposal has been a distressing issue of the State since decades. Conventionally, the disposal of solid waste is by incineration wherein two types of ashes generate, the bottom ash and the fly ash. Overcrowding and traffic congestion in cities and towns have grown to be alarming. There is an urgent need for upgradation of the public infrastructure, and improvement in sewage and sanitation system.

Decadal growth of urban population in the State varies between 39 and 48% indicating the rate of incremental growth of the urban centres. The construction sector has grown from about 2.5% in 2014–15 to nearly 30% in the following year, decreasing to 9.4% during 2016–17. This growth rate generates nearly 0.2 Mt of construction and demolition waste annually. Utilization of waste materials and by-products is a partial solution to environmental and ecological problems, since it not only reduces costs, but also has indirect benefits such as reduction in landfill cost, energy saving, preventing potential pollution among others.

Keywords Solid waste · Plastic waste · Fly ash · Construction waste · Bio-medical waste · Eutrophication

6.1 Introduction

Municipal solid waste (MSW) is defined as the waste collected by the municipality that includes residential, industrial, institutional, commercial, municipal and, construction and demolition (CDW) waste (Hoornweg et al. 2015). The quantum of waste generated globally in 2012 for instance was 1.3 Gt, while it is projected to increase to about 4.3 Gt (Hoornweg and Bhanda-Tata 2012) by 2025. The MSW is a major pollutant of the environment. Nature, type and quantum of solid waste generated have varied over the years. Increased climate challenges and the scarcity of

A. G. Dessai, *Environment, Resources and Sustainable Tourism*,
Advances in Geographical and Environmental Sciences,
https://doi.org/10.1007/978-981-99-1843-0_6

virgin natural resources in construction projects have moved the research momentum towards sustainable economy. The conventional economy model is centred on manufacture, use and dispose. The current trend, however, is to transform, protect and generate resource, revenue and materials back into circulation, thereby becoming eco-friendly and cost-effective. More attention needs to be directed to energy conservation, environmental protection and resource reuse by recycling and even reduction in use, that form the social benefits of a responsible society contributing to the overall sustainability of the region.

In recent years, the State of Goa has adopted several steps for systematic collection and segregation of waste especially from cities and towns. Despite these efforts, a large number of dumps of waste is still subjected to unauthorized, improper, uncontrolled and illegal dumping by the roadside or along highways or in open spaces or in water bodies and sometimes in open drains and even on the pavements, posing a serious threat to the environment.

6.2 Municipal Solid Waste: Disposal Concerns

The MSW comprises the unwanted or useless solid materials generated from human activities in the residential, industrial or commercial areas, commonly referred to as trash and garbage. It consists of day-to-day items such as product packaging, clothing, furniture, bottles, food scraps, newspapers, appliances, batteries, among several other items of daily use. The MSW may be categorized in three ways:

- Origin: domestic, industrial, commercial, construction or institutional
- Contents: organic material, glass, metal, plastics, paper
- Hazard Potential: toxic, non-toxic, flammable, radioactive, infectious and the like

The abbreviation MSW emphasizes the importance of sustainable material management (SMM) which refers to the use and reuse of materials in the most productive and sustainable way across the entire life cycle. The SMM practices to conserve resource, reduce waste, minimize the impact of various materials on the environment and moreover, help in slowing down the process of climate change.

6.2.1 Solid Waste Management Policy Framework

Prior to 2016, the responsibility of waste management was incorporated within the Solid Waste Management Rules (Shekhar et al. 2017). However, in later years the Ministry of Environment, Forest and Climate Change modified the rules to regulate the management of waste, in particular, the CDW Waste Management Rules of 2016. As per this policy, institutions have been allocated different responsibilities (ibid.). The State governments are assigned a key role in organizing CDW waste management within the areas of their jurisdiction.

6.2.2 Solid Waste Management of the State

Background-Site for Waste Disposal: Several attempts by the State to identify a suitable permanent site for disposal of garbage have been objected to by the local civic bodies and the populace at large. This is a problem specific to Goa that has more to do with scarcity of land, poor incentives, lack of proper awareness and misconception about garbage disposal. The general perception among locals is, land allotted for garbage disposal is a misuse, as either it does not fetch or fetches least monetary benefits. Experience of temporary garbage sites elsewhere in the State, such as the one at Sada (Mormugão) and Sonsoddo (Margao) has been bad, due to air pollution (stench from decomposition of organic matter), dispersal of waste and nuisance from stray animals and birds in the vicinity of dumps. Hence, the attitude 'not in my backyard' which hinders the State from finding a suitable site. The solution appears to be in educating the public on garbage disposal and how it is more valuable than commercial price of land. It is necessary also to provide better incentives by way of price or compensation for such land allocations.

As per the State Pollution Control Board, the quantum of solid waste (excluding the construction- and electronic/electrical-waste) generated in municipal areas is about 190 t per day. An additional 200 t is contributed from rural areas. The quantum of waste is likely to increase in future considering the proposed developmental plans of the State viz. Panaji Smart City Project, and Island-development Project and many others. Largely, solid waste, which at present includes large quantity of construction waste, could be classified into (i) biodegradable and (ii) non-biodegradable, each about 50% of the total. The tourism policy stipulates separate collection and aggregation of the two categories of waste, however, this is rarely, if ever, followed in practice, except in some municipal areas in very recent times.

6.2.3 Pollution from Leachates

In common parlance a leachate is a fluid percolating through the solid waste dump-site or landfill, generated from moisture/water present in the waste and that from outside, including rainwater filtered through the waste. Scientifically, it is a by-product of physical, chemical and biological decomposition of municipal solid waste, usually formed at MSW dump sites, landfills, incineration plants, composting plants among others with a high level of toxicity (Zhao 2018). A leachate may be defined as any contaminated fluid that is formed from water percolating through a solid waste disposal site, accumulating contaminants that percolate into the subsurface (Cheremisinoff 1997). The leachates originate due to the presence of high moisture content of the disposed waste. As the waste is compacted or as it chemically reacts, the bound water is released in the form of a 'leachate'.

There being no barrier to control the movement of leachates, they migrate freely and contaminate surface and subsurface waters and the soils through which they percolate. Pollution from leachates is a matter of global concern both for reduction of the source and removal of the pollutants. With time, the landfills age and readily degradable organic components undergo anaerobic degradation, as such older landfills are more stabilized and may form less organic material. Hence, leachate generation is maximum during the first 5–10 years and further it declines with age.

The quality of the leachate varies depending upon the waste composition, water budget and the physical, chemical and biological conditions at the dump site. As most dump sites by and large contain high organic waste, anaerobic biological decomposition during the early life of the dump, say first 10 years, determines the leachate quality, which is dominated by organic carbon. These processes then influence the pH and the mobility of other compounds from the waste containing iron, calcium, chromium, cadmium among others that are released due to bio-chemical decomposition. The leachate pollution is usually high over a fairly long period of time presumably over decades or longer (Ehrig and Stegmann 2018). The leachate composition could be considered under four categories:

- Dissolved organic matter
- Inorganic macro components
- Heavy metals
- Xenobiotic organic compounds

In order to control the spread of effluents from the site, the use of drainage wells through the dump layers is proposed. These structures are made of fixed metallic cylinders (~1 m diameter) made of metallic net, filled with gravel. Such drainage wells can also be combined with biogas extraction facilities (Cossu 2018).

6.2.4 *Products of Waste Incineration*

The disposal of solid waste is usually by incineration. Two widely used processes include the refuse-derived fuel (DRF) process and mass burning process (Siddiqui 2008). The by-product of incineration process is ash. It is typically about 1–30% by wet weight and 5–15% by volume of the wet MSW, depending upon the nature of the incineration plant. Two types of ashes are produced, the bottom ash and fly ash. The bottom-ash residues are large and are removed from the bottom of the incinerator, whereas the fly ash residues are fine-grained particles entrained in the exhaust gases. At times the fly ash is combined with bottom ash into a mixture, which is then disposed off. The latter constitutes about 75–80% of the total combined ash stream. The ash consists primarily of glass, ceramics, ferrous and non-ferrous metals and minerals. It also contains toxic metals such as lead and cadmium along with organic compounds such as dioxins, also called ‘dirty dozen’ which is a group of highly toxic chemicals known as persistent organic pollutants (POPs). They accumulate in fatty tissue of animals and have a half-life of 7 to 11 years in the body. MSW fly ash could be used as partial replacement for cement in concrete mixes as a supplementary cementitious material (refer Sect. 6.2.7). Ceramics, glass and glass ceramics are high consumers of silicate-based natural raw materials.

6.2.5 *Biodegradable Waste*

As of now, the disposal of biodegradable waste in some municipal areas and starred hotels is attempted through composting of solid waste. However, in the countryside and along the beachside shacks and huts, the two types of waste is not segregated separately and dumped together. It is either buried in pits contaminating the groundwater (refer Sect. 5.3.1) or burnt in open polluting the air.

Waste Disposal* via *Waste to Energy Conversion: Recently, the State has set up a biomethanation plants at Saligao (since 2016) to produce electricity from biodegradable waste. In this process, organic waste is microbially converted to biogas under anaerobic conditions. The gas is then utilized to produce electricity. The Saligao plant has the capacity to treat 100–250 t of wet waste and produce 7.0 MW of electricity of which 4.0 MW is consumed by the plant while the remaining is distributed through the grid. The plant collects mixed waste from about 25 towns and villages. The plants of smaller capacity exist at Cuncolim, Margao, Bicholim and Mormugão. A 100 t/day capacity solid waste management facility is under construction at Cacora (Curchorem, Quepem taluka) whereas a 250 t/day capacity plant is proposed to be set up at Bainguinim (13 km from Panaji). These facilities will recycle, sort, segregate, bio-methanate and transform the garbage and biodegradable trash into organic material on the lines of the Saligao facility or with the potential as a fertilizer and cater to South Goa and, Tiswadi taluka and surrounding areas, respectively.

The non-biodegradable, component of the garbage which was almost non-existent earlier, is a prime concern now. Of late, it is collected at Karapur Sarvan (Bicholim taluka) and Cacora.

6.2.6 *Biomedical Waste*

The State has 148 hospitals and nursing homes in addition to 462 clinics and dispensaries, 43 veterinary institutions and 101 pathological laboratories (Anon, 2021). The biomedical waste generated by the hospitals and nursing homes amounts to 1.271 t/day. As of now the State does not have a functional common biomedical waste treatment facility. The one under construction was likely to be operational by 2021–22. The waste is largely incinerated, there may be instances when it is dumped with other solid waste exposing the waste-workers and the public to a health and environmental hazard.

The liquid chemical waste need to be segregated at the source, pre-treated and neutralized prior to mixing with other effluents for disposal. The waste being infectious and biohazardous need to be handled cautiously and transported separately for disposal. The Goa Medical College and hospital, in the absence of a centralized biomedical waste treatment facility, disposes the waste through an autoclave/incinerator set up about two years ago. It has a capacity to incinerate waste at a rate of 100 kg/hour. The establishment of a common biomedical waste treatment facility at Kundaim, is being contemplated by the State.

6.2.7 Sustainable Management of CDW

India, in general, and Goa in particular, are witnessing a construction boom due to fast-paced urbanization which is leading to increased stress on exploitation of primary earth resource materials used for construction purposes. Among the locally procured resources road-metal aggregate (crushed stones), sand, clay are the main resources used in this sector on which there are several restrictions and ban on exploitation in the State due to which there is irregular supply of material and price rise. High demand and lack of regular supply of material is also encouraging considerable illegal activity in exploitation, as is regularly reported by the print and electronic media (*Herald* dtd. 18. 8. 2022). As a consequence of short supply and increased costs of materials, promotion of secondary materials, as substitutes for the primary, is possibly the only option available.

A serious aspect of construction activity is the building waste generated and its disposal. The waste includes bricks, concrete, wood, insulation materials, nails, electrical wiring, rebars (mesh of steel bars/wires) as well as waste originating from site preparation, such as dredging material, tree stumps and rubble (Siddiqui 2008). This waste is collectively referred to as 'construction and demolition waste' (CDW). It amounts to about 500 t per day of which the largest component about 30% is bricks and mortar aggregate, about 20% concrete, almost identical is cement mortar, the balance making up the rest. It impacts the environment at every step, right from extraction of raw material, processing, manufacturing, transportation, construction and disposal of material at the end of building's functional life. In fact, with increasing awareness of environment, scarcity of landfill spaces and their ever-increasing costs, utilization of construction waste has become an attractive alternative to disposal and a profitable venture. Large quantum of waste and its impact on the physical environment requires adequate solutions. Despite construction regulations (Goa Land Development and Construction Regulations 2010), the waste is quite often dumped on public roads. Utilization of waste materials and by-products is a partial solution to environmental and ecological problems. It not only reduces costs but also has indirect benefits such as reduction in landfill cost, energy saving, avoidance of potential pollution among others (*ibid.*).

***CDW Waste Management System*:** CDW waste management system includes systematic collection and classification of waste material and segregation from the source, read place of waste generation, systematic transport, storage, followed by suitable processing either to recycle or reuse materials that of value with least rejection, which invariably goes as landfill.

Recycling and Reuse of Construction Waste: New innovations turn waste into valuable supply of resource to the construction industry. The composition characteristics of the waste decides its suitability either in recycling or reuse. A detailed use of various classes of materials is available in Shekhar et al. (2017). A brief outline is provided here. The concrete waste is generally recycled to generate aggregate of different sizes which are then utilized as replacement for natural aggregates in the construction work.

The crushed aggregate of appropriate size (9.5–37.5 mm, broadly up to 20 mm) class made from a mixture of CDW is termed as recycled aggregate. It can be used to the extent of about 20–30% along with the natural aggregate for the construction of non-load bearing structures (ibid.). In recent times, it is more commonly used in the manufacture of pre-fabricated, moulded blocks such as pavers and so forth. The granular waste is utilized for road construction. The crushed hard material of sand size (2.0–4.75 mm) grade is generally reused as replacement for natural sand. Sand is one of the main ingredients of concrete used for plastering, concreting and finishing works for all type of constructions. The fraction less than 2.0 mm is classified as fines of which the fraction less than 0.75 mm, referred to as dust, is generally used as landfill (ibid). However, as the latter contain cement as one of the constituents, it migrates into the substratum and pollutes the soil and groundwater aquifers.

Green Concrete as Recyclable Resource: The recycled concrete materials, fly ash find application in the manufacture of Green Concrete in road engineering. Considering the shortage of stone aggregate, waste concrete can be processed into concrete aggregate, thereby saving on transportation as well as material cost of stone aggregate. It is an innovative approach to utilize construction waste in road engineering construction. The Chinese have been using this technology for sustainable development of basic transportation facilities (Ma et al. 2022).

The scrap metal, steel, aluminium, copper is subjected to smelting operations and recycled to manufacture new products. The materials can also be reused. Likewise, the undamaged wood work such as wooden frames, doors, panels as also unbroken bricks are sold for reuse in secondary open market.

Polymer Concrete: An important aspect of sustainable management involves the planning for infrastructure facilities both in terms of establishing infrastructure, operation and maintenance that essentially involves the durability aspect of structures. Just as recycling of waste is a way of economizing on costs, in a similar manner, use of quality materials also greatly helps in conserving resources. As far as Goa is concerned, it is a coastal region where the level of humidity is high throughout the year and to add to that, the air is charged with sodium chloride due to the proximity to the sea. Most reinforced cement concrete (RCC) structures are subjected to reinforcement corrosion, spalling and micro-cracks formation due to the ingress of moisture and chlorides.

Research in construction materials since the fifties has developed innovative materials such as polymer concrete which have high compressive strength, fast rate of curing, and are resistant to corrosion. Polymer concrete is a type of concrete in which polymers are used as binding materials instead of Portland cement. When the polymer contains the ester functional group as linkage in their main chain, it is called a polyester and the resultant product is polyester concrete. (In common parlance a Polyester is a kind of plastic usually derived from petroleum. Dacron, widely used in the manufacture of textiles in the sixties and seventies, is one of the examples of polyester polymer. Polyester concrete is the name given to a composite that in effect is *not a concrete* in the strictest sense.) If a polymer is used in addition to cement, then it is known as the Polymer Modified Concrete. Most commonly used polymers are Furan resins, Acrylics and Styrene-Acrylics, Polyvinyl Acetate, Methyl Metacrylate, Epoxy Resins, Poly Urethane Resins and so forth. The aggregate used, i.e. silica plus granite plus limestone plus others, should be of good quality, free from dust and dry, failing which the desired bond strength is not achieved. This material has high thermal stability, is resistant to variety of chemicals and is successful in both new constructions and repair work. Due to its great adhesion, it is preferred for repairs of both polymer concrete structures and conventional cement concrete structures. Since it has low permeability and corrosion resistance, it is ideal for use in swimming pools, nuclear power plants, marine works, water tanks, sewerage structures, particularly repairs of manholes due to invincibility to toxic and corrosive sewage gases and bacterial attack. The main shortcoming of polymer concrete is its high cost compared to the conventional concrete, however, recent progress has led to significant reduction in cost making its use more widespread.

6.2.8 Pozzolans in Light Weight Concrete

The pozzolans consist of a group of silicious and aluminous materials which in finely divided form and in presence of water, reacts chemically with calcium hydroxide [$Ca(OH)_2$] at ordinary temperature to form compounds possessing cementitious properties (Siddique 2008). This supplementary cementing material requires that kaolinite is roasted, but not burnt, which results in the formation of mullite that is not reactive.

The kaolinite [$Al_2Si_2O_5(OH)_4$] clays, which are traditionally used in the manufacture of porcelain, find application as supplementary cementing materials especially in producing inorganic geopolymer cement. Calcination of kaolinite clays leads to their dehydroxylation [loss of $(OH)^-$] between 500–800 °C (e.g. Mackenzie 1957; Smykatz-Kloss 1974). The heat treatment of kaolinite leads to thermal activation of the mineral and the process is known as calcining. It results in partial breakdown of crystal lattice and the formation of a transitional phase with high reactivity, known as Metakaolin (Siddique 2008). This is a pozzolanic material.

Portland cement with activated pozzolanic binders finds applications in Geopolymer Concrete which is made by reacting aluminate and silicate bearing materials with a caustic activator such as fly ash or iron slag. It has lower emissions of CO_2 to the atmosphere.

Potential Applications: The metakaolin is a silica-based product that on reaction with $Ca(OH)_2$ produces CSH (calcium silicate hydrate which is the primary product of hydration of Portland cement and is primarily responsible for imparting strength to cement-based materials) gel at ambient temperature (*ibid.*). Metakaolins find applications in high strength lightweight concrete, fibre cement and ferro-cement products, glass fibre reinforced concrete, mortars, pool plasters, to improve finish, colour and appearance. It increases compressive strength, resistance to chemical attack, reduces shrinkage due to particle packing and makes concrete denser. In the course of mining for iron, manganese and bauxite often large quantities of kaolinitic clays are excavated as overburden. If stacked separately, these could be used in the manufacture of pozzolans. The Industrial Policy of Goa, 2003 stipulates conversion of industrial waste into useful products. Attempts were made on a pilot scale to make bricks from such overburden material in the past.

The reuse of raw materials is being increasingly promoted in recent times to conserve and preserve the environment, to reduce stress on natural resources and to control resource consumption. Recycling has the potential to generate employment by creating new industrial ventures, thereby contributing to the overall growth and the economy of the region. Thus, efficient use of secondary resources too can have beneficial effect on the society.

Alternative construction materials are many that can be used effectively in concrete for instance, crumbed rubber, steel slag, copper slag, glass concrete, saw dust, ceramic waste, quarry dust among others.

6.2.9 Electronic Waste

In recent years electronic waste (e-waste) such as discarded computers, televisions, VCRs, stereos, copiers, fax machines, PCBs, catalytic converters, various types of telephones among others, in addition, have joined the list of solid waste about which not much information in the State exists. The quantum of e-waste in the country has been increasing from 0.77 Mt in 2017–18 to 1.04 Mt in 2019–20. The State generated electronic waste on an average about 3 t /per month during 2019–20 (Anon 2021). Combined electrical and electronic waste averaged about 15 t/month. The e-waste requires a different treatment of disposal and a part of it can be recycled too and is usually treated separately from the other types of solid waste. In recent years e-waste is utilized for the extraction of gold. The National Policy on Electronics 2019 and National Manufacturing Policy 2011 also stipulate sustainable e-waste management including safe disposal and recycling, and creation of manufacturing zones that are clean and energy-efficient. A State policy is desirable for the management of e-waste in future.

6.2.10 Plastic Waste

There are forty-four manufacturing units of plastic bags. The manufacture of plastic sheets less than 50 microns in thickness is banned by the State. The quantum of plastics consumed annually has been growing steadily. Plastics used in packaging, automotive and industrial applications, medical delivery systems, artificial implants, other healthcare applications, water desalination, land/soil conservation, preservation and distribution of food, housing, communication materials, security systems and for various other uses contribute significantly to the solid waste management problems. Primarily due to its large visibility, plastic waste has been viewed as a serious solid waste problem. In the plastic waste, polyethylene forms the largest fraction which is followed by PET (polyethylene terephthalate) of which plastic bottles are made and are widely used for soft drink packaging. It is the most common thermoplastic polymer resin of the polyester family used as fibre in clothing, containers for liquids and foods and several other applications. The microplastic pellets constitute a serious threat to marine biota (Veerasingam et al. 2016). The recycled polymer material as a substitute for aggregate in concrete has been studied and it is recommended that recycled polymer modified blocks and concrete can be used in non-load bearing structures, floating structures and where light weight material is desirable (Rahman and Islam 2012).

Need for Recycling: The plastic waste generated in 60 cities of the country is estimated to be over 15,342 t per day. The State generates about 23 t of plastic waste per day, in which 20 t is from urban centres. About 30% of it is non-recyclable. A plastic waste processing plant is proposed to be set up at Pernem. It would have a capacity to treat 10 t of plastic daily and generate about 6.7 t of furnace oil, 3 t of metallurgical coke and 500 kg of wax from the non-biodegradable plastic component of the garbage. The viability of this plant/process will have to be ascertained. The plastic consumer industrial units should be taken on board in launching plastic recycling programmes. As of now, the local bodies such as the village *panchayats* are entrusted with the responsibility of collection, storage, transportation, processing and disposal of plastic waste. However, cleaning of the surroundings including beaches is entrusted with the tourism department which in turn allots the cleaning contract to some private agency. The public roads, beaches and even open spaces and water bodies are the most polluted. The State has initiated efforts of segregating plastic waste. In most cases due to lack of appropriate infrastructure, the waste used to be burnt in the open, polluting the air. However, recently it is being segregated separately and is sent to cement factory in Karnataka for co-incineration.

Managing Plastic Waste: Alternatives to the use of plastics is one of the ideal solutions to reduce the menace of plastics. Other steps could be the cost of recovery and recycling of the plastic product should be inclusive in the retail pricing of the product, as is done in some western countries. Encouragement to the production of biodegradable plastics is yet another initiative. Some regulatory mechanism to control marine and coastal pollution from tourism activities should be in place. Awareness of pollution from plastic waste and systematic collection for disposal/recycling is most important, both for the public, especially for the small restaurant owners, tourist boat operators, local (Indian) tourists and others to be more sensitive towards minimizing the plastic waste/litter in particular and the litter in general.

Plastic waste is being tried as a substitute of aggregates in concrete. Shred and ground plastic waste is mixed with cement and fly ash to form plastic concrete. It is highly versatile, light weight, durable concrete, resistant to water and chemicals. The electrical and thermal insulation properties are excellent and is comparatively less expensive.

Disadvantages: Experiments have shown that the bonding capacity of plastic is less, hence the flexural, tensile and compressive strength of concrete is less; also, the melting point of plastic concrete is low.

6.2.11 Glass Waste

Use of glass products has increased considerably resulting in large quantum of waste glass. Glass is a transparent material formed by melting of a mixture of materials such as soda ash (Na_2CO_3), silica, calcium carbonate at high temperature followed by cooling during which solidification occurs without crystallization (e.g. Siddiqui 2008). There are two types of waste glasses: coloured and colourless. The latter is readily recycled. Coloured waste glass with low recycling potential is generally dumped as landfill. Since it is non-biodegradable, landfill disposal is not an environment friendly solution. In recent years, however, glass is utilized as road construction aggregate asphalt paving, concrete aggregates, glass tiles and bricks, wall panels, fibre glass insulation, glass fibre, abrasive, agricultural fertilizer, landscaping, tableware, hydraulic cement and several other applications. Recycling of glass from municipal solid waste for use as raw material in new glass products is limited due to high cost of collection and processing (hand sorting), and specifications that limit impurities (e.g. ceramic, ferrous metal, paper, plastics among others) in the glass production process. There is considerable interest in the development of concrete from waste glass referred to as 'glasscrete'. Waste glass is crushed into specified size and can be replaced by fine aggregate in concrete, a 30% replacement provides 9% higher compressive strength compared to conventional concrete (Mary et al. 2020). It has better workability compared to that containing natural sand, increased durability as compared to conventional concrete and improves the overall efficiency of the concrete.

Disadvantages: One of the significant disadvantages with glasscrete is alkali-silica reaction. The silica in glass chemically reacts with the hydroxyl ion from cement and forms silica gel which absorbs water leading to cracks in the cement. Besides, the cleaning process is quite expensive, organic residue on glasscrete harms the strength of the concrete.

6.2.12 *Tyre-Rubber Waste*

Significant quantities of scrap tyres are generated every year. Over the years, disposal of tyres has become one of the serious problems in environment protection. Landfilling presents difficulties due to the scarcity of free land-availability for disposal.

Waste rubber tyre when crushed is called crumb rubber. Innovative solutions for tyre rubber disposal include (i) use of tyre-rubber in asphaltic concrete, (ii) incineration of tyres for production of steam and (iii) reuse of ground tyre-rubber in a number of plastic and rubber products. A 5% replacement in concrete by crumb rubber provides 5% higher compressive strength compared to the conventional concrete (Mary et al. 2020). The scrap tyres are also used as fuel for cement kilns, as feedstock for making carbon black, which is used as reinforcing filler in tyres and other rubber products. It is also used as a pigment in plastics, paints and inks.

Scrap tyres can be managed as a whole-tyre, as slit-tyre, as shredded or chopped-tyre, as ground rubber or as crumb rubber product. In addition to recycling into new products, ground rubber is used in floor mats, carpet padding, mud-guards and in road asphalt. Crumb rubber is used as an asphalt binder modifier in hot-mix asphalt pavements. A considerable proportion of scrap tyres goes for retreading. Automobile and truck tyres are increasingly used in Portland cement concrete mortar.

6.3 Industrial Pollution: Concern of Effluents

Background: In order to discourage large manufacturing industries (Secondary Industries) that normally lead to excessive pollution, the State resolved to encourage non-polluting eco-friendly industries, save the mining industry, the Primary Industry, which generates raw materials, and the most polluting of them all, could not be immediately discontinued as it shouldered the State economy. Accordingly, the Goa Industrial Development Corporation was established. It has facilitated the creation of several industrial estates/parks in Goa especially to attract ‘environment-friendly’ (Tertiary and Quaternary) industries. However, a number of pharmaceutical industries (not discussed here) which are considered to be strong polluters of the environment and the aquatic ecology in particular, were also permitted. The negative impact of the production of pharmaceutical products on the natural environment is well-known and remains largely unregulated. Though, the industry is export-driven, generates foreign exchange and contributes about 10% to the national economy, it is vulnerable to the local populace and the natural resources.

The Tertiary sector which includes the Service Sector in particular, generates nearly 63% of the global wealth. There exist 23 functional industrial estates which house 283 factories and 4987 small-scale enterprises (Anon 1997) of which 541 units generate hazardous waste; these data do not include the additional 15 industrial estates that are proposed.

These industrial estates were primarily intended to provide employment to the locals, however, in most industries the locals employed do not average to even 10%. It is necessary to undertake the cost–benefit analysis and ascertain whether the benefits to the State are commensurate with the land and the subsidies, inclusive of the tax holiday that the State provides to the concerned industry. It is well-known that in most cases the industries close down after a few years and the State has to spend additional funds on litigation, mainly for land recovery. Consequently, the land remains idle for several years and is a loss to the State exchequer.

Hazardous Waste: The industrial estates are spread over different parts of the State. The GSPCB has identified 1628 industrial units that generate hazardous waste in the State (Anon 2021). Most of these are located on tablelands (*plateaux*) situated about 20–100 m above MSL. In many cases, due to the laxity in implementations of statutory provisions, industries with potential to cause pollution have been also permitted. Some of the industrial units in the Kundaim, Cuncolim, Pilerne and some others too, have the potential to release toxic effluents into the streams and rivers. For example, there are two units in Cuncolim estate which are engaged in the production of metals Cu, Zn, Co and Ni from imported raw material, the by-product during these operations is the solid hazardous waste. The surrounding well waters (refer Sect. 5.3.2) show high concentrations of Ni, Fe, Mn and Cd, and Pb beyond permissible limits of potable water (Anon 2014). The total quantity of hazardous waste generated in Goa is 36,105 t, of which about 76.12 t is of landfill quality, 12,015 t is incinerable, 3599 t is recyclable and 12,879 t is utilizable (Anon, 2021). It is essential that a hazardous waste treatment, storage and disposal facility is created by the State on an urgent basis. A hazardous waste treatment plant is proposed at Dharbandora, however, as in all other matters, here too land availability has been a matter of concern.

6.3.1 Type of Effluents

The effluents from the industrial estates, in particular the distilleries, are the major contributors of pollutants to the river systems. In addition, the chemical factories and fertilizer plant also discharge their effluents in the rivers. In order to efficiently manage the effluents, they need to be treated by means of an Effluent Treatment Plant (ETP). It is a process design for treating the effluent wastewater for its reuse or safe disposal, so as not to adversely affect the environment. The effluent in general could be considered under three heads:

- Influent: Untreated industrial waste water
- Effluent: Treated industrial waste water
- Sludge: Solid part separated from wastewater by ETP

The objectives of the ETP facility are many, of which the following are the most significant:

- to clean the industry effluents and recycle them for further use
- to reduce the usage of fresh/potable water in industries
- to lessen the cost on expenditure for water procurement
- to meet the standard of emission or discharge of environmental pollutants
- to safeguard the environment against pollution and contribute in sustainable development

Type of Pollution: The pollution caused by the increase in the concentration of chemical nutrients is known as ***eutrophication***. It is generally indicated by the increased trend of blue green algae. Depending on the degree of eutrophication, subsequent negative environmental effects such as anoxia, and severe reduction in water quality, fish and other animal populations may occur. Fish mortality has been noticed in several streams and rivers due to the industrial effluents. One of the severest of cases to be cited was that which caused massive fish mortality in the Zuari estuary due to the discharge of effluents from the fertilizer plant of the Zuari Agro Chemical Limited in 1979. Other instances of industrial pollution are also discussed (refer Sect. 5.3.2).

In general, tourism is grossly instrumental in polluting land and sea, destroying wild life and natural habitat, and, degrading and depleting energy and natural resources.

6.4 Overstressed Land Resources

Real Estate Development-Fallouts: Real estate here includes essentially property consisting of land and buildings such as residential, commercial and industrial complexes. Availability of land being severely stressed, the land prices are soaring and as such mindless construction activity has grossly increased. This is largely due to migration of locals to cities for employment and education, disintegration of joint families and current trend of nuclear families, foreign immigrants, influx of employment seekers from other states, among others. Construction of infrastructure for hospitality, health care, recreation, education, other public utility services, industrial units among others, has been on the rise. This has fomented widespread and unprecedented encroachment on private and public lands and illegal land deals.

6.4.1 Growth Pattern

The extent of growth of construction in towns and cities can be appreciated from the rate of growth of urban population. In 1960, 14.8% of the total population was urban population, in 2001, it was 49% and presently it is 52%. Decadal growth of urban population varies between 39 and 48%. It was a record 132% during 1960–1971. This provides a measure of the incremental growth to the urban centres. The growth of the construction sector has varied from about 2.5% in 2014–15 to nearly 30% in the following year, decreasing to 9.4% during 2016–17 (Town and Country Planning Department, Govt. of Goa). However, from the occupancy of the residential spaces (66%), it is apparent that there is a mismatch between planning and requirement, as more than 20% of the built-up spaces remain vacant.

6.4.2 Density of Structures

Considering the demand for housing and commercial requirements, and constraint of land availability, the 'set-back' (which is the distance between structures and between the road and structure) permitted in urban areas varies from 3–8 m, based on the bye laws applicable at the place in question or based on the width of the road. Moreover, little provision is made for consolidated open spaces and city commons. Despite lesser set-back, elevation of 8 storey is permitted (set-back to building-height ratio is substantial). This provides a measure of the density of structures in urban settlements. Similarly, the 'floor space index' (FSI) permitted in urban areas is 2, which means the total permissible construction-area is twice or more than that of the actual area of the plot of land. The FSI for residential areas is 1, meaning permitted construction is equal to the area of the plot of land.

Limitations of Increased FSI: In urban residential areas, parking is an associated problem. Due to high FSI and low set-back, stilt-parking should have been made compulsory, to control road congestion. However, considering the scarcity of land availability, particularly for commercial purposes, and the high returns involved, stilt-parking in residential cum commercial establishments is not made compulsory, and neither underground nor overhead parking. With the result, there is crowding of vehicles on roads in city centres during peak hours. Moreover, in emergencies, such as fires, floods, earthquakes or medical emergencies, it presents considerable difficulties.

Cost of power, water and even construction material has increased manifold. Affordable housing, particularly for the low-income groups, is provided by the State through the Goa Housing Board. Liberal financing policy in the housing sector has also encouraged the construction industry.

6.4.3 Unaffordable Land Prices

The coastal *talukas* of Bardez, Tiswadi, Mormugão and Salcete are the most populous accounting for 23% of the total area, where 58% of the total population of the State lives. The land prices in the towns and cities have skyrocketed. Land prices in remote villages in the southern and eastern parts of Goa vary between ₹2,000 to ₹3,000 per m^2 whereas in the capital city, the prices vary anywhere between ₹5,00,000 and

₹20,00,000 per m^2. Housing particularly in urban areas is becoming unaffordable for sustainable development.

References

Anon (1997) Masterplan for development of groundwater in Goa state. Central Groundwater Board, Ministry of Water Resources, Government of India, Bangalore, Report No. 3 http://cgwb.gov.in

Anon (2014) Annual Report for the year 2013–14. Goa State Pollution Control Boards, 1–256

Anon (2021) Annual Report for the year 2020–21. Goa State Pollution Control Boards, 1–354

Cheremisinoff NP (1997) Treating contaminated groundwater and leachate. Groundwater remediation and treatment technologies. http://www.elsevier.com

Cossu R (2018) Physical landfill barriers: Principles and engineering. Solid waste landfill, concepts, processes and technologies, 271–287. org/https://doi.org/10.1016/B978-0-12-407721-8.00015-2

Ehrig H-J, Stegmann R (2018) Leachate quality. Solid waste landfilling, concepts, processes and technologies, 511–539. https://doi.org/10.1016/B978-081551411-4.50009-0

Hoornweg D, Bhanda-Tata P (2012) What a waste: a global review of solid waste management. Urban development Series, Knowledge Paper No 15, World Bank, Washington DC http://openknowledge.worldbank.org/handle/10986/17388

Hoornweg D, Bhanda P, Kennedy C (2014) Peak Waste: When is it likely to occur? J Ind Ecol 19:117–128

Ma Y, Ji Y, Jin L (2022) Green and environmental protection recycled concrete in road engineering. Scientific Programming, vol. 2022, Article ID 5377984.https://doi.org/10.1155/2022/5377984

Mackenzie RC (1957) Thermal methods. The differential thermal investigations of clays. Mineralogical Soc, London, pp 1–22

Marry JA, Gobinath R, Shyamala G, Chary KR (2020) Waste products as an alternative construction material- A Review. IOP Conference Series: Material Science Engineering 981. https://doi.org/10.1088/1757-899X/981/3/032086

Rahman MdM, Islam MdA (2012) Recycled polymer materials as aggregates for concrete and blocks. J Chem Eng 27:53–57

Siddiqui R (2008) Waste Materials and By-products in Concrete. Springer-Verlag, Berlin-Heidelberg, p 413

Shekhar AR, Chandra K, Rathi V (2017) Training manual on construction and demolition waste management in India for cities and towns. http://reurban-industrial.in/live/hrdpmpmaster/igep/e64918/e67075/e67087/GIZ_OD_eTrainingManual.pdf

Smykatz W (1974) Differential thermal analysis, applications and results in mineralogy. Springer Verlag, Berlin- Heiderberg, p 185

Veerasingam S, Saha M, Suneel V, Vethamony P, Rodrigues AC, Battacharya S, Naik BG (2016) Characteristics, seasonal distribution and surface degradation features of microplastic pellets along the Goa coast, India. Chemosphere 159:496–505

Zhao Y (2018) Pollution control technologies for leachate from municipal solid waste, from environmental materials and waste. http://www.sciencedirect.com

Chapter 7
Sustainable Tourism

Abstract Tourism is viewed as a useful instrument to foster development, as it generates employment, creates wealth and its rapid growth is considered as an attractive possibility for sustainable development. However, mass tourism and rapid development in Goa have impacted the ecology, inflicted enormous stress on the environment, energy and socio-economic conditions of the local community. The positive effects have been shielded by the adverse ones largely due to the disconnect between the policy on the one hand and its execution and outcome on the other.

Management of the coastal zone has been a serious concern both due to population pressures and preferred site for urbanization and settlement consequently the mangrove forests have borne the brunt. Large-scale migration to cities and immigration have impacted housing, transportation, health care, education and all other activities. Loss of local traditions, native customs and destruction of local culture show a rising trend.

Although tourism was intended to be inclusive by integrating all stakeholders, particularly the women, underprivileged and economically weaker sections of the society, the benefits of economic development have not permeated the society as expected in the nation's goal of achieving 'Aatma Nirbhar Bharat' (self-reliant India). The State should be vigilant enough to curb tourism that jeopardizes the safety of women and children, and control growing crime and antisocial activities. Tourism activity should be regulated to sustainable levels by determining the carrying capacity of the State.

The programs and strategies of the tourism industry in the State are critically analysed and based on the human, economic, environmental and social sustainability of the State resources, the future perspective of sustainable tourism is outlined as a means of sustainable development.

Keywords Sustainable tourism · Mangroves forests · Air emissions · Sewage and sanitation · Community-based tourism · Low carbon footprint

A. G. Dessai, *Environment, Resources and Sustainable Tourism*,
Advances in Geographical and Environmental Sciences,
https://doi.org/10.1007/978-981-99-1843-0_7

7.1 Introduction

Tourism is one of the fastest-growing sectors of the global economy. It is a prime revenue earner in many parts of the developing world. However, it is a complex enterprise in which economic and political factors combine with those of geographical and recreational nature (Hall 1998). Obviously therefore, the definition of the tourism policy varies depending upon what the governments choose to project or not to project as regards the industry (Hall and Jenkins 1995). Most studies dealing with tourism tend to focus on the achievements of tourism rather than on the impacts on the physical and socio-economic imbalances that it brings (op. cit.). The difficulties encountered in the developmental activities and the impact on the natural resources contrast sharply with the praise heaped by the governments on tourism's contribution to economic development (Jenkins 1980; Lamb 1998; Hillali 2007).

Tourism has been a major industry operational in Goa since the later part of the twentieth century and has been undoubtedly the backbone of the State economy, especially during the post-mining ban period. As such, this chapter focuses on the Tourism industry, which, despite being a revenue earner, has been a formidable consumer of the State's natural resources, at quite a rapid pace. Since the impact is on resources, it has an influence on the physical and socio-cultural environment, and the lifestyles of the people of the State. Hence, it has been felt necessary to critically analyse tourism in the State, and how it could be made a sustainable activity to foster sustainable development. In addition, the preservation and protection of the environment, the conservation of earth resources and preservation of local culture are of vital concern for this State considering its size, limited resource base and a rich cultural heritage.

7.2 Overview of Tourism Industry

Tourism is the fourth largest industry in the world economy next only to Manufacturing, Agriculture, and Construction and Technology. It has gained importance since it generates employment and helps in poverty alleviation, while also facilitating regional infrastructure development and promoting local art and craft. However, the impact of tourism could be either beneficial or detrimental to the society. Although tourism invariably contributes to the national economy and helps in balance of payment, mass tourism can have catastrophic repercussions on the ecology and inflict enormous pressure on environment, energy and other natural resources, while impacting the society and local culture (Lea 1988).

The third world countries, particularly the emerging economies, endeavour to cash in on this industry in an attempt to boost foreign investments and financial resources. However, integrated tourism planning is inadequate; thereby there is a disconnect between the policy, on the one hand, and its execution and outcome, on the other. Inefficient resource management and the need for a robust regulatory mechanism are needed in the third world countries, to check the spawning of unregulated growth of this industry, often resulting in environmental and societal distress. Furthermore, the involvement of local community needs to be encouraged instead of exclusively focusing on high-end tourism, so that the benefits accrued from Tourism percolate down to the weak and underprivileged sections of the local community. Broadly, the geodynamics of the environment has a direct bearing on the sociodynamics of the region. The socially engineered changes in the natural habitat not only affect the natural ecology but also the social ecology. Hence, any discussion on environment cannot overlook its social dimension that is, the impact on the society and the people at large.

It is, therefore, proposed to review the experiences and strategies of the tourism industry in the State during the last forty years with a view to determining the potential role of tourism in managing the State economy and its impact on other sectors, such as the resources and the physical and socio-economic environment. The economic, environmental and social sustainability of State resources has been empirically determined for the purpose from the perspective of the local community. Based on this information, it is intended to outline the future perspective of sustainable tourism in the State and formulate a set of recommendations that could be taken into consideration by the governmental agencies, professionals, planners, policy-makers and other stakeholders to devise strategies that could be explored to promote tourism in the State as a means to sustainable development.

7.2.1 Definition of Sustainable Tourism

Tourism in a broad sense is a service sector which requires a range of goods and services for its support and generates revenue as foreign exchange. It is a multi-sector economy that requires the involvement of other sectors such as agriculture, banking, insurance, transportation, food, culture, other technical services and products required to support both travel and tourism. Being an employment- and income-generating enterprise, it is an industry that has great appeal in the third world countries despite its shortcomings. Hence, a lot of attention is devoted to the systematic study of this industry so as to make tourism as profitable as possible with least damage to the environment and the society at large, for as long a time as possible. Here lie the roots of sustainable development of this enterprise.

The concept of 'sustainable development' has been introduced in Chapter 3 (refer Sect. 3.6). As detailed in the Brundtland (1987) Report 'Sustainable development is the ***development that meets the needs of the present (people) without compromising the ability of future generation to meet their own needs***'. Humans, in addition to natural resources, also require social and economic resources for their sustenance. Hence, social and economic development forms an integral part of sustainability. In this holistic approach, ecological, social, economic and human dimensions are considered together in an attempt to find lasting prosperity. Thus, sustainable development inter alia includes and is dependent on four distinct areas—the four pillars of sustainability, namely:

- Environmental Sustainability,
- Economic Sustainability,
- Social Sustainability (UN General Assembly 2005) and
- Human Sustainability.

The environmental sustainability involves maintaining a balance of all environmental systems while the natural resources within them are consumed at a rate where they are able to replenish themselves. The economic sustainability envisages that human communities across the globe are able to maintain their independence and have access to the financial and other resources that they require to meet their needs. The social sustainability stipulates that universal human rights and basic necessities are available to all people. They have access to sufficient resources to maintain their families and communities healthy and secure where personal, labour and cultural rights are respected and where people are free from discrimination. The human sustainability is aimed at maintaining and improving the human capital in society and is concerned with the health and education systems, access to services, nutrition, knowledge and skills, all of which come under the umbrella of human sustainability. In the business world, sustainable development can be applied to corporate policy, where it encompasses three key areas, namely the environment, economics and society, that are informally referred to as the planet, profit and people.

The concept of sustainable tourism has its roots in the concept of sustainable development. Several definitions of 'Sustainable Tourism' are available in literature, however, the one of the World Tourism Organization (WTO, also referred to as OMT—Organización Mundial del Turismo 1997) could be the one most suitable for our purpose. As per this definition Sustainable Tourism can be defined as ***'tourism which leads to management of all resources in such a way that the economic, social and aesthetic needs can be fulfilled while maintaining cultural integrity, essential ecological processes, biological diversity and life support systems'***. In other words, tourism can be a useful instrument of the planners and policy-makers to encourage development; however, it must be implicit that it should be first of all well understood and organized, since its positive aspects at times, get shielded by the adverse impacts, as has been observed in certain instances (Zappino 2005). Sustainable tourism is one that meets the needs of the tourists, the tourism industry and host communities without compromising the ability of future generations to meet their own needs. According to the WTO sustainable tourism should:

- make optimal use of environmental resources maintaining essential ecological processes and helping to conserve natural heritage and biodiversity;
- respect the socio-cultural authenticity of the host communities, conserve their cultural heritage and traditional values, and contribute to inter-cultural understanding and tolerance;

- ensure viable, long-term economic operations, providing socio-economic benefits to all stakeholders, including stable employment and social services to host communities, contributing to poverty alleviation.

In summing up sustainable tourism is a complex issue that requires the study of the present status of tourism, impact on economy, biodiversity, environment and the socio-cultural life of the region. It is imperative that to maintain sustainability, tourism should be economically viable, friendly, both environmentally and ecologically, and acceptable socially and culturally, failing which the sustainability merely remains on paper. It is, therefore, necessary that we as a State assess, implement and improve the sustainability measures that we follow and the manner in which we operate our tourism business. These aspects on a broader canvass are coupled with the overall sustainable development of the region.

7.2.2 *Sustainable Development Index*

The fast pace of development and urbanization witnessed in India in general, and Goa in particular, has placed high demands on each and every type of resource and on almost every component of the economy. In order to quantify the level of sustainability achieved by a country, anthropologist Jason Hickel (2020) has devised a Sustainability Development Index (SDI). The SDI measures the ecological efficiency of human development by recognizing that the development must be achieved within the confines of the planetary systems. The index is calculated by dividing the nations 'human development score' obtained from statistics of life expectancy, health and education, by its 'ecological overshoots', the extent to which the per capita carbon footprint exceeds the Earth's natural limits.

Carbon Footprint: A carbon footprint is the total amount of greenhouse gases including carbon dioxide and methane that are generated by our day-to-day actions (nature.org). Globally, the average carbon footprint on a per capita basis is about 4 t, while that of United States is 16 t. It is estimated that the average global footprint per capita needs to be brought to about <2 t by 2050, in order to avoid a 2° C rise in global temperature. India's total emissions from energy use in 2018 were 2.25 Gt of CO_2 in which contribution from power generation was 53% whereas emissions from the industry contributed 25%. Based on these emissions in the environmental health and ecosystem vitality ranking, India is ranked 178 out of 180 countries. Under the climate change mitigation category, India's ranking is 165 out of 180. The nation's current CO_2 emissions (2021) are 2.88 Gt with a major contribution from the transportation sector, the result of burning of fossil fuels. China has the highest carbon footprint at 29.18%, followed by United States at 14.02%, India at 7.09% and Russia at 4.65%. On a per capita basis China's footprint is 7.38, United States is 15.52, India 1.91 while among the developed countries such as Denmark the per capita footprint is 6.65, Norway is 8.28 and Sweden is 4.54, whereas among the

Caribbean islands, where tourism is their major revenue earner, Cuba leads with a lowest footprint of 2.68.

The SDI index enables us to bring compatibility between development of a nation and its ecological well-being. It provides us a rationale to make normative judgements about countries and help them improve their environmental strategies. It is possible to compare nations with one another as far as the achievement of the level of sustainability is concerned. It also gives indirectly a handle to gauge the level of concern of a nation for the humanity at large, and the planet as a whole. Accordingly, countries with strong human development and a low environment impact, score highly, whereas countries with poor life expectancies and literacy rates with higher ecological limits tend to score low in terms of this index.

In a list of SDIs of 165 countries for the year 2019, India figures at 52nd with an SDI of 0.696 which it shares with Venezuela, whereas another developing country Brazil (SDI: 0.747) ranks 33rd. Costa Rica tops the table (SDI: 0.850) followed by Sri Lanka (SDI: 0.836). No country, however, reaches an index of over 0.9 (sustainabled evelopmentindex.org). A small country, such as Cuba despite sanctions from United States and UK for over six decades, is among countries at the top of the list with a score of 0.859 (Trinder 2020). The state of Kerala has shown a smart performance in this listing with a SDI score of 0.803. It is a measure of the state's achievement with a relatively high human development in terms of life expectancy, education and income, and its consumption-based CO_2 emissions and material footprints which are within planetary boundaries.

7.2.3 State Tourism Policy

As this narrative is primarily concerned with sustainable tourism, it would be apt to outline the tourism policy of the State and to present a few thoughts on the policy framework of this sector. As noted before, tourism is a recent phenomenon, which caters essentially to the development of one sector namely the economic sector. It is obvious from the literature survey that economic aspect of tourism has the precedence over other sectors as far as most governments are concerned (Velasco, 2018) and particularly so, of the emerging economies. The state of Goa is not much different in adopting a similar approach, wherein a high-end tourism is envisioned. This is also reflected in the tourism policy of the State which almost treats the tourism policy not much removed from the economic policy. However, when viewed from a larger perspective, tourism policy includes much more than economic policy and imposes a profound influence on the society. Therefore, defining the tourism policy has several complexities which need to be addressed in any generalized definition. In a broad-based definition, tourism policy is a set of dialogues, decisions and practices adopted by governments either in consultation with the social sector or without, in order to achieve multiple objectives related to tourism (e.g. Velasco 2018).

When seen from an academic point of view, there are too many complexities involved in simplifying the definition of tourism, since it is an industry that transgresses several sectors of the society. The diverse fields of knowledge stress certain aspects of the issue that tend to overshadow others. Being a multifaceted industry, it is almost impossible to frame a definition that would satisfy and be universally acceptable to one and all. On the contrary, a definition that makes sense to each one from the field related to the analysis of the issue, is a satisfactory definition, as long as it is sufficient for the purpose for which it is proposed. Tourism is rapidly gaining importance in almost all disciplines, primarily, since it involves economies of nations which form the basis of livelihood and well-being of society at large.

7.2.4 Principles and Objectives of the State Policy

The vision of the State policy is to '***make Goa the most preferred destination around the year for high spending tourists in India by 2024, and a world class tourism destination by 2030***' (goa.gov.in). The document prepared by the government in consultation with the industry and other stakeholders enunciates the sustainability, Goa's uniqueness, safety, ease of mobility, global visibility, authenticity of Goan heritage, lifestyle and culture, and redevelopment of the State as a diversified tourist destination.

The objectives of the policy include (i) attraction of a mix of domestic and international tourists, (ii) creation of an environmentally friendly and socially sustainable destination with balanced development of the coastal regions and the hinterland, (iii) improvement of infrastructure including accommodation, connectivity, power, water, waste treatment among other services, (iv) provide a world-class experience with a distinct Goan identity, diversity and heritage and (v) incorporation of modern technology to enhance the quality of tourism products.

The governance and the institutional framework will broadly lie with the Goa Tourism Board, Goa Tourism Development Corporation and the Directorate of Tourism.

For a State constrained by an inadequate resource base, the new tourism policy is quite likely to usher in a socio-economic reorganization impacting several sectors of the economy and leading to a palpable shift in the sustainability model. For an efficient management of this sector in the light of new policy, an assessment and evaluation of State resources is desirable, since it will involve not only conservation of heritage, biodiversity and the like, but will also entail providing stable employment and income-earning opportunities towards poverty alleviation among several other aspects. It is likely to further stress the land use, already under considerable stress. The rise in real estate demand will cause an increase in building cost and land value, which will adversely impact the local population, since increase in income is not proportionate. Similarly, increasing demand for basic services and goods from the tourists will cause price rise again affecting the locals. Large-scale migration to the cities and to the State is likely to be a major concern, unless that is contained, the propagating slums will continue to grow in major cities and the gains of development will not be truly realized. The disparity in the developmental divide between the western region of the State and the eastern hinterland may become still sharper. A study to determine the carrying capacity would be able to provide a measure of the quantum of load and resource consumption that it can sustain, as well as the quantity of waste likely to be thus generated. These studies will enable the State to manipulate, so as to keep the numbers below the maximum carrying capacity lest the system does not collapse. If this exercise is dovetailed with vulnerability study, it will facilitate protection of the resource from degradation. These studies will help decide the course of efficient resource management, facilitate mid-course correction if need be, and aid the process of resilience building against potential threats, since some of the changes expected are likely to be of irreversible nature.

7.3 Tourism: The Case of Goa

Tourism industry, although well recognized as an engine of economic growth in general, is most beneficial to the local populace when the region in question produces at least some consumables in excess of its own requirements. Unfortunately, this is not the case in Goa where every consumable item needs to be procured from neighbouring states (the so-called leakage effect) in order to meet the requirements of the locals as well as the tourists. There being a mismatch between the demand and supply, the prices of foodstuffs shoot up. Locals end up paying the same or even more than the tourists (hotels can procure their requirements at wholesale rates and pass on some of the benefit to their customers). Consequently, there is an increase in the cost of living which has a cascading effect on many social parameters. Wages need to be necessarily increased leading to increased labour costs; cost of production increases and the product fails to be competitive in the market, leading to decreased profits. This discourages the setting up of labour-intensive industries employing local labour. The decline of agriculture in the State is a case in point.

Despite active involvement of a fairly large chunk of the local community in the tourism sector, the earnings are not sustainable. Two examples should suffice to illustrate the point: (i) a Goan publisher intending to print his book/magazine would rather do so in neighbouring towns such as Belgaum or Kolhapur since, even considering the time and money spent on travelling, his cost would be less than half of what he would spend to get the same done in Goa, (ii) the labour from Karnataka, employed on construction projects, bring their victuals (foodgrains, condiments among others) from their native place, live in makeshift shanties and forage for firewood; thus their earnings become their savings, which can fetch much more when repatriated home. Hence, they can afford to work for wages that would be uneconomical (below subsistence) to a local. This is also the reason why many regular employers (such as hotels and even small-scale industries) prefer outstation labour over locals (Borges, per.com.). Thus, although tourism generates employment, it also leads to increased local unemployment and outmigration, an aspect camouflaged by the larger beneficial effects of the industry. The unregulated growth of both construction and tourism sectors in the State has led to massive consumption of materials impacting the overall sustainability of the tourism industry. The local community has to bear the brunt of high prices of foodstuff, housing, maintenance among others (Menon 1993; Sawkar et al. 1998; Venugopal 2021).

7.3.1 Demographic Pressures on Resources

For the last forty years, Goa has been an important tourist destination both for the international and domestic tourism. In fact, it has acted as a gateway to many holiday destinations within the country and to other countries in the Indian Ocean. Approximately 5–6 million tourists (0.15–0.25 million international) on an average, both domestic and international, visit the State annually. The details of tourist arrivals (Table 7.1) suggest a progressive increase in arrivals year after year indicating the preference of tourists for Goa, notwithstanding the fact that the industry in the State has weathered the storm of several periods of recession including the Gulf War (1990–1991), the Kargil War (1999), the Indian Parliament Terrorist Attack (2001), the Mumbai Terrorist Attack (2008) and the ongoing global Coronavirus (COVID-19) pandemic, first reported from China (31 December 2019). The present bed-strength (all categories) of the State is about 0.102 M. Despite the perturbations, tourism is the prime industry on which a major part of the economy of the State is primarily dependant. During the fiscal 2020–2021 the State earned ₹825.02 billion (US$11.14 billion) contributing 14% to the foreign exchange earnings of the country. The positive impact on the financial resources is due to its large employment-generating capacity and its multiplier effect on the economy. Additionally, locals have been benefitted from improvement in infrastructure, such as the road infrastructure, bridges, telecommunication, electricity and water supply, healthcare facilities among others.

However, mass tourism and rapid development have had collateral adverse effects on the environment, ecology and socio-economic condition of the community. The influx of tourists and concentration of tourism-related activities at specific destinations and at specific times (October to December) has generated tremendous stress both on the resources and the local community. The stress is likely to increase in future with the commissioning of the new airport at Mopa, which would handle 30 M passengers per year. In addition, changes and at places, destruction of natural resources, cultural and ecological heritage have been profound.

The seasonality of the tourism activity also impacts the resources; they remain largely unutilized in the rainy season (June to September) whereas during the tourist season they are overstretched due to excessive floating population (nearly double the resident population), beyond the carrying capacity of the region (Venugopal 2021). The unskilled workforce required by the industry is impacted due to unemployment during the lean season. The employment pattern has come to depend on tourism,

Table 7.1 Statistics of Tourist arrivals and Revenue collection in Goa during the period 2011–2019

Year	Total tourists M	International M	Charter M	Revenue (Domestic) M₹	Revenue (International) M₹
2011–2012	2.68	0.48	0.169006	15.0	169,006
2012–2013	2.63	0.49	0.215304	–	–
2013–2014	3.54	0.51	0.261452	–	–
2014–2015	4.76	0.54	0.154047	1.85	0.2077
2015–2016	5.65	0.68	0.158779	2.65	0.2697
2016–2017	6.9	0.89	0.233	3.08	0.33
2017–2018	7.08	0.93	0.2259707	3.38	0.3459
2018–2019	7.1	0.90	0.218	–	–

as a large chunk of the local community has moved away from their traditional occupation, namely agriculture and related activities; some have shifted to tourism centres. The infrastructure and the recreational facilities developed for tourism have not benefitted the local community as most of the labour force is from neighbouring states. The new tourism products (recreational facilities) too are at times outsourced to companies from other states. This has resulted in increase in the cost of goods and services, increase in labour cost, expensive housing, consequently affecting the cost of living as mentioned above. In more ways than one tourism pressures have adversely affected the physical- and socio-economic environment, and the moral and social fabric of the State (refer Sects. 7.7–7.9). Most issues pertain to corporate social responsibility, which, in effect, revolves around 'sustainable development' and how it can be managed to meet the requirements of present generation without compromising the ability of future generations to meet their own demands. Sustainable tourism, therefore, has been a valuable tool to alleviate the adverse impact of tourism and to accomplish sustainable development.

7.3.2 *Sustainable Tourism: Potential for State Initiatives*

According to Euromonitor International (http://www.euromonitor.com), the Scandinavian countries lead the way for sustainable tourism. Sweden and Finland alternately top the list of ten well-managed countries followed by Austria, Norway among few other European countries. The reason is their tourism development has taken due care of all the four pillars of sustainability, without concentrating merely on economics, as usually is the case with many developing economies.

As the State intends to embark on a sustainable tourism policy, planning and management of tourism will have to be approached in a holistic manner by addressing, environmental, financial and local community concerns. Needless to add, economic considerations always overweigh, yet environmental concerns should not be sidelined, especially at a time when massive infrastructure development is in progress, strongly impacting the environment and ecology in various ways. The intent of protecting the ecology need not always be sabotaging the very initiatives but mending ways to protect and preserve the natural diversity with least damage and destruction, simultaneously exploring ways for revitalization and rejuvenation of the ecology. The other aspect of critical importance to sustainability is permeation of the welfare and benefits of the tourism activity to the local community. This will not only benefit the local stakeholders but will be able to project to the world community, that a crucial aspect of sustainable tourism is not overlooked by the State from implementation.

The State will have to espouse fresh initiatives by adopting a twin approach of modern and traditional development initiatives. Goa as a state produces no significant power of its own, neither hydel nor thermal. It has ample potential and opportunity to invest in the development of renewable energy, both solar and wind energy. Many of the desired factors for green power generation are ably conducive. The wave energy system, although could be cheaper, is still in early stages of development, making speculations on development and maintenance costs difficult. Likewise, recycling of some commodities, for instance reclaiming of used water and rainwater could be yet another effective initiative that could considerably reduce the stress on the groundwater resources. The other potential area where more aggressive efforts could be directed is reuse of waste, especially the biodegradable waste, that could be utilized to generate, gas, heat and/or electricity. Some initiatives in this direction are already underway, these need be scaled up to commercial and industrial level power generation. These modern technologies could be concurrently supported by revival of some of the traditional occupations, such as traditional pottery, carpentry, making of leather goods, bamboo-craft and cane-craft among others, that are on the verge of extinction in the State or are being practised by migrants from neighbouring states. These and similar initiatives could provide a dual flavour to the State economy at the same time, helping preserve traditional occupations and providing both employment and active participation of the local community in the tourism enterprise. The contribution to nonconventional energy will be able to reduce carbon footprints, an aspect of crucial importance to the State.

The other aspect of sustainability involves the reduction of threats and risks involved in safeguarding both the local community and the tourists. Among many of the developed nations of the world, Saudi Arabia, although ranks 72nd in the list of the world's most sustainable tourism destinations, it ranks first in the risk mitigation strategies, including geopolitics, natural disasters, man-made catastrophes and diseases. Goa could take a leaf out of their policy in areas relevant to the State, for instance in adopting suitable measures in curbing pollution-related respiratory diseases, waterborne diseases, reduction in the rate of road accidents among others. Better preparedness from the State in counterterrorism, and in tackling international criminals, drug menace, flesh trade, child trafficking among others, would be desirable.

7.3.3 Coastal Resource System and Tourism Products

The tourism activity in the State is largely confined to the coastal zone (beach tourism) which forms one of the important components of the environment. Management of this region for sustainable use has become a critically important issue in recent years, due to increasing population pressures as preferred sites for settlement and urbanization, and the resultant increase in demand on the coastal zone. These factors present considerable challenges to sustainable exploitation of resources in the coastal zone.

Coastal zone, in a broad sense, may be defined as 'that space in which terrestrial environment influences marine environment and vice versa'. It is an area of transition where terrestrial and marine environments interact to form unique conditions. Coastal zone, therefore, embraces inshore waters, inter-tidal areas and extensive tracts of neighbouring land. It includes a wide variety of diverse habitats and ecosystems such as estuaries, mangrove swamps, creeks, lagoons, (Fig. 7.1) bays among others. The areas of real interest and concern are regions where human activities are linked with both land and marine environments, and could be referred to as '***coastal resource systems***'.

In contrast to the traditional idea of general tourism, the modern concept of tourism stipulates different types of tourism, for instance beach tourism, sports tourism,

Fig. 7.1 A seasonal lagoon at Ambaul, popularly known as Cola beach, in South Goa

ecotourism to quote a few examples, are different tourism products (Zappino 2005). The State has developed a few of these products laying emphasis primarily on natural assets, sun-sea-sand (sea beach) and to some extent eco-tourism which are the main products on which the industry depends. Other product such as cultural tourism, heritage tourism, medical tourism, wellness tourism, wildlife tourism, film tourism, adventure tourism and a few others have a good potential and although introduced, these need to be given further impetus. During the last four decades, the region focused heavily on mass tourism directed primarily on the beaches and being the most popular destinations for recreation, this type of tourism is likely to be developed further exerting additional stress on the beaches.

Need to Overcome Inhibitions of Free Trade: Tourism being a broad sector requiring services of several other sectors and a wide range of simple and complex goods and services to support, it should foster development of better infrastructure. It should also develop partnerships with other countries thereby augmenting the competitiveness of the local stakeholders. The region also should get actively involved with international organizations such as WTO, FTAA, EPA, which could have significant positive impact in achieving sustainable tourism development in this region. The potential of developing such partnerships is being increasingly recognized by the local policy-makers. In this context, some of the past actions of the country do not appear to be in keeping with the desired strategies. For instance, three years ago, India withdrew from the Regional Comprehensive Economic Partnership (RCEP), a free trade agreement among 15 nations including China. Apart from RCEP, the other two agreements are the Comprehensive and Progressive Agreement for Trans-Pacific Partnership (CPTPP) and the Indo-Pacific Economic Framework (IPEF). India's apprehensions in withdrawing participation from RCEP were due to the fear of Chinese goods entering duty-free in India, flooding the home market. However, in practice it is observed that our trade volume with China has increased for the past three years. Likewise, despite withdrawing from RCEP, the country has aggressively followed free trade with some member countries such as Australia, UK, UAE, Canada and EU.

In a world slowing down due to recession, even if the share of trade goes up by 3–4%, it would greatly help by providing a boost to the Indian economy. India's growth depends on being globally competitive in both manufacturing and services. It is favourable for the country to discontinue protective measures to safeguard the domestic industry from global competition. Since employment generation is a necessity, the country can benefit from global engagements. It is learned from past experience that labour-intensive exports provide us an edge, may it be in textiles, tourism, agro-processing or software services. It is in country's interest to join free trade across all sectors, thereby providing much needed boost to the economy, tourism included.

7.4 Impact of Tourism: Stressed Environment

The boost to tourism has indeed provided new avenues for employment in the hospitality and services, transport and communications, and all other allied sectors. As mentioned earlier, it is the fourth largest industry in global economy next to Manufacturing, Agriculture, Construction and Technology. It has brought prosperity to the locals. However, the undesirable effects of tourism have been unprecedented, for instance:

- overexploitation and damage to the terrestrial and marine resources,
- uncontrolled growth of infrastructure, encroachment on private and public land, illegal construction in towns and coastal stretches
- overcrowding in cities and towns,
- increased waste, excessive pollution and augmented emissions of different types,
- significant impact on local culture and a strong possibility of increase in social evils such as increase in crime, and antisocial activities,
- loss of local traditions and native customs and
- destruction of culture due to copying of lifestyles of tourists, are the other shortcomings that impact the society.

7.4.1 Distress on Land Resources and Mangroves

Considering the size of the State and its demography the land available for various purposes is clearly inadequate. The eastern part of the State is largely hilly and thickly forested. Large stretches of it are identified as reserved forests, wildlife sanctuaries, bird sanctuaries among others. The remaining portion constitutes the mining belt with small stretches of narrow valley-floors that form the agricultural farmlands (Fig. 7.2) (refer Sect. 5.2.3).

Land use and Infrastructure: The western part of the State, in particular, the coastal belt is thickly populated and is the most stressed. Most developmental activities, such as hotels, housing complexes, recreational parks, other infrastructure projects, are all confined to this narrow, about 75 km long stretch. The strip of land, less than a km wide from the beach, is the most attractive and overstressed part of the coastal belt. There is a ban on construction within 500 m from the high tide line (CRZ Notification 1991, 2011). This line was intended more to keep the beaches and the estuarine stretches free from infrastructure encroachments; however, the statutory provision was conveniently interpreted to be applicable to marine (beach) environments and consequently the estuaries were spared of this restriction. In the case of inland waters, the ban was restricted to 200 m and is likely to be reduced further to 50 m. Despite the ban several housing complexes and hotels have sprung up within the restricted zone, right at the waterfront (Fig. 7.3) threshold with utter disregard to the stipulated norms. The built-up within the 500 m buffer from HTL (high tide line) has increased from 19% in 1991 to as much as 40% in 2019 (*Hindustan Times*

Fig. 7.2 Agricultural land on tidal flat (*khazan*) festooned by mangroves, Zuari estuary at Rachol, South Goa

dtd. 18.9.2021). Earlier such violations were more common in the coastal belts of Bardez and Pernem *talukas*, however, in recent years even the remote beaches such the Cola beach in Canacona, is not spared. In fact the preference today is more for such exclusive, remote hideouts. The overall situation is quite distressing due to loss of farmlands, open spaces, biodiversity and forests. Simultaneously, there is an, adverse impact on earth resources viz. sand and gravel, laterite, groundwater and land in general leading to overuse of resources and causing pollution, damage (sewage disposal, waste mismanagement) and destruction of ecology.

Physical Damage to Marine Resources: Tourism in the State is heavily concentrated along the beaches. They are the most vulnerable and overcrowded during major part of the year (Fig. 7.4). They are stressed both due to anthropogenic activity and nature's fury.

Threatened Mangrove Forests: Mangroves occupy a small area of about 2,000 ha but constitute a complex agro-pisciculture-saltpan ecosystem that is the ***most threatened***. Of the total mangrove cover, about 80% occurs in the Mandovi, Zuari, the Cumbarjua canal, Sal (Fig. 7.5) estuary among others (Untawale et al. 1973). These wetlands cover an area of about 18,000 ha (Anon 2021) locally known as '*khazans*', under tidal influence, have been reclaimed for centuries with a system of *bunds* and *sluice gates* to control the flow of tidal waters. They are significant for the ecosystem and biodiversity of the region. The flora consists of about 15 species (of family Rhizophoraceae, genus Rhizophora) of which *Rhizophora mucronata*, *Avicennia alba*, *Avicennia officinalis*, *Sonnerecia alba* and *Sonnerecia officinalis* predominate (Jagtap 1986).

Fig. 7.3 Permanent concrete structures right at the waterfront, north Goa

Fig. 7.4 A view of the overcrowded beach at Baga on New Year eve (Courtesy: R. Naik)

These wetlands, as mentioned earlier, being the most productive natural habitats host a variety of biotic communities and hence constitute ecologically sensitive areas of the State. The wetlands could be of varied types depending upon the climatic conditions, topography, hydrology, water chemistry, soil types, biotic communities, among other factors. However, broadly, two types of wetlands are recognized: (i) coastal or tidal wetlands and (ii) in-land or non-tidal wetlands. The wetlands from the State could be categorized as tidal wetlands, since they are not completely isolated from tidal influence, thereby generating environments of varied salinities. The State has identified six of these areas, namely the Xeldem and Nanda *lake*s in Cacora, Toyyar *lake* in Chimbel, Dashi *lake* in Revora, Sarzora *lake* in Chinchinim and Cottambi *lake* in Quepem as official wetlands as per the central government Conservation and Management Rules 2017.

Fig. 7.5 Luxuriant mangroves on the western banks of the Sal estuary at Cavelossim, South Goa

Urbanization activity in the townships largely takes place at the expense of mangrove forests. For example, the infrastructure development projects in the capital city Panaji are implemented solely by reclaiming the mangroves and encroaching on *khazan* lands, there being no other free land available (Fig. 7.6). The Kadamba Transport Corporation bus terminus and the Patto Plaza Complex occupy erstwhile mangrove areas. The new Mandovi bridge, Atal Setu and its approach roads have further impacted the already shrinking mangroves cover. A portion of this ecosystem has been seriously impacted due to the non-availability of fresh saline waters as a consequence of the construction of approach roads to the bridge, resulting in the death of large chunk of mangrove forest. Apart from economic considerations, mangroves serve as a protective barrier during storms, floods, sea-surges and other similar events, thereby controlling the wave energy and eroding capacity of waters.

In the stretch between Agacaim and Siridao, the *khazan* lands are being utilized for infrastructure projects such as construction of housing complexes. Large infrastructure projects have come up at the waterfront threshold (Fig. 7.7). These are just a few examples to illustrate the point. Similar activities are observed all over Goa. The State has now been more cautious in sanctioning new projects. A number of housing projects are being carefully scrutinized by the State authorities primarily due to encroachment on *khazan* lands and sand dunes.

7.5 Congested Road Traffic: Remedial Measures

There are over 1.4 M registered vehicles in the State at the end of the fiscal year 2019.

Fig. 7.6 The aerial view of lush mangroves (background) at the Kadamba Transport Corporation bus terminus, Panaji encroached upon by urbanization activity (Courtesy: C. Barreto)

Fig. 7.7 Large infrastructure project at the waterfront in North Goa

7.5.1 Increased Emissions

Overcrowding and traffic congestion in the cities and towns has grown to be quite alarming, largely related to migration of population to cities and rapid development of new settlements as discussed above. This has implications due to air emissions, noise pollution, solid waste and littering, release of fumes, heating due to emissions, increased use of lights and LED signs (light pollution), traffic congestion among others. In a population of about 14.59 lakhs, there are about 600,000 vehicles out of which about 420,000 are two wheelers. This ratio of nearly 4:1 is quite high for any state by Indian standards. According to National Family Health Survey, 45.2% of the Goan families own a car while 86.7% households have a two wheeler (Zee News dtd. 22.05.2022, zeenews.india.com). In comparison, the survey states that barely 7.5% of Indian families own a four wheeler and 49.7% own a two wheeler.

7.5.2 Growth in Road Accidents

The road network has not changed substantially and neither has the State widened the existing roads (save developments underway) required to cater to the voluminous increase in traffic. It is quite worrisome that a small State has amongst the highest accident rate in the country. The number of fatal road accidents is on the rise: 270 in the year 2005, 303 in 2006, 322 in 2007 and 200 in 2008. The number of accidents increased to 3881 in 2013, 4304 in 2016, 3440 in 2019, 2275 in 2020. Although the numbers show a declining trend, the severity was higher than in earlier years with 213 fatalities and 150 grievous injuries. The overall rate of accidents in the State is highest in the country with 109.4 accidents per one lakh population, this is about four times the national average (27.6/one lakh population) (*Times of India* dtd. 06.01.2015, 21.04.2015, and 27.05.2022). Majority of road accidents involve two wheelers. Given the size of the State (3702 km^2) and the population density of 394 per km^2 (North Goa: 471 per km^2, South Goa: 326 per km^2), these numbers are alarming.

7.5.3 Challenges to Change

The transport lobby in the State has always been a force to reckon with. The State transport came into existence nearly two decades after liberation. Even after nearly thirty years of liberation, the public transport in the State was not nationalized due to strong influence of the transport lobby on the government of the day. Finally, during one of the short stints of the President's rule over the State, the governor of the State took a wise and bold decision and the State transport was nationalized, albeit on certain trunk routes. The same is true as regards the use of fare metres for public transport vehicles, such as taxis, three-wheeler autos and others. Although as per the statutory provisions all vehicles are fitted with fare metres, they are not being used due to the high-handed attitude of the transport lobby. It influences the government not to make the use of fare metres compulsory, leading to considerable hardship to the locals and specially the tourists. Motor-cycle taxis (locally known as *motor-cycle pilots*), a mode of public transport unique to the State, is also without fare metres. In an international tourist destination such as Goa, hiring of a taxi of one's own choice is often precluded, due to a strong control by the local taxi operators.

7.5.4 Improvement of Infrastructure

Improvement of the road network by constructing parallel and ring roads, by-pass roads to towns, flyovers and subways, widening of roads is an urgent need. The total length of the road network was 5,119.58 km in 2011 and has not increased substantially over the years. It is unable to cope up with the phenomenal rise in traffic. Enhancing the efforts at educating vehicle drivers on traffic safety measures as also construction of pavements for the pedestrians is essential. Economic development has brought in a change in lifestyle. Drivers prefer to pay a penalty for the traffic offence rather than abide by the road safety instructions. Self-discipline and civic sense, that used to be the hallmark of the society in the pre-liberation era, have been severely eroded in recent times. There is an urgent need to improve the road infrastructure, construct multi-lane highways, enhance the traffic regulatory system, reduce traffic congestion by creation of freight logistic hubs, make adequate provision for parking of vehicles, both in housing complexes and in public parking spaces/lots, and lay emphasis on the public transport system which would alleviate the stress on the road traffic. Simultaneously alternative means of transport such as the development of waterways and suburban railways should be also attempted. There is a need for better coordination between modes of passenger transport: air, road, rail and water to enable passengers to switch the mode at lesser hardship and expenditure. The prime requirement, however, is to have a blueprint for the road traffic of the State for a projection of at least two decades.

7.6 Construction Industry

The construction sector, in general, is one of the pillars of the economy, the builder of modern India. However, the workers who create the base of the new economy, live in a time warp, trapped in low skills, low pay and insecure working conditions. About a third of the workforce is women and even children. The women in particular face serious problems related to work such as wage discrimination, gender and sexual harassment, unhealthy job relationship, lower wages and the like. Despite this the industry attracts female workers.

7.6.1 Real Estate and the Workforce

Goa being the most preferred holiday destination has ample scope for real estate and property investments. North Goa promises a good return on investment whereas South Goa could be the ideal choice for spending quiet, quality time. The construction sector in the State is a developing sector and in the last decade with a couple of exceptions, particularly in 2004, it has witnessed a boom especially in major towns such as Panaji, Margao, Vasco da Gama, Mapusa and the villages of the coastal belt. Value-added construction in Goa increased from ₹14.03 billion in 2012 to ₹24.17 billion in 2021 growing at an average rate of 6.65% (knoema.com). The industry attracts a huge workforce, largely labour force, which is often exposed to job and income insecurity, exploitation, violation of rights and absence of effective legal protection. The low-paid migrants although are beneficial to the State as cheaper workforce, they also tend to strain the State resources. Hence, the low-economic-impact migration to some extent is not completely stress-free to the State economy.

Workforce is the prime concern of the construction sector and of the State. Project management in construction involves efficient utilization of labour, material and equipment. Material handling includes procurement, inventory, shop fabrication and field servicing that require special attention for cost reduction. Labour productivity in construction is a crucial component, often defined as output per labour hour. Since labour forms large part of the construction activity and the quantum of labour hours in performing a task is more susceptible to the influence of management than are materials or capital, the productivity measure is referred to as labour productivity.

Construction labour is usually of three types, unskilled, skilled and specialized. The majority of labour is unskilled and helps in masonry work. The skilled labourers are masons, carpenters and others, whereas the specialized labour is the one doing the task of flooring, marble work and so on. Most labour in the State is procured from the neighbouring states. The total workforce, in the organized and unorganized sectors according to the Labor Department of the State, has a strength of nearly 3.0 lakhs, up from 1.55 lakhs in 1991. These include the migrants and locals which form a small percentage.

7.6.2 Construction Materials and Substitutes

It is possible to make an assessment of the demand for construction materials in the State from the consumption of cement utilized by the industry. The State utilizes about 0.4–0.5 Mt of cement per year with an increment of nearly 10%. A realistic estimate would suggest that the requirement of cement in 2022 would be over 0.5 Mt which would require about similar quantum of stone aggregate and more than 1.5–2.0 Mt of sand. Considering the quantum of construction activity in the State, the demand for the availability of virgin material for the construction sector is tremendous, in relation to the size of the State. Although for RCC structures, steel and cement are procured from other states, the aggregate, sand, laterite and water are indigenous. If excavated/exploited to the required levels, good quality resources may not last for a couple of decades considering the limited resource base. Therefore, in order to conserve the State resources, it is essential that part of the material at least is recycled and part is substituted by other suitable materials. For instance, natural sand is being substituted by crushed slag. Likewise, laterite bricks could be suitably substituted by bricks made of crushed slag, fly ash and a suitable binder such as cement. The fire-clay bricks also form a good, time-tested substitute, considering the availability of clays both from virgin areas and some mine dumps (where stacked separately) which may require some blending. Such attempts will one, lessen the pressure on land resources used as dumping sites which leads to contamination of other resources and two, it will control the proliferation of the construction and demolition waste.

7.7 Socio-Cultural Conflicts

Background: Goa is one of the more developed states within the Indian Union. As a noticeably secular and civilized state, it stands for values such as coexistence, plurality and inclusiveness. Moreover, it represents a way of life promoting religious as well as cultural freedom. It is a tiny region within our vast country, which displays a healthy blend of Indian and Western culture and has a population that largely believes in peaceful and harmonious coexistence. No wonder, that the State has attracted a large number of people from various parts of the world who have immigrated and settled largely along the coastal stretch. According to an assessment, the State is threatened by reduction of native population due to migration from other states. There are apprehensions that by 2030 the migrant population may outnumber the natives diluting the ethnic character of the State. This has reflected in a huge boom in construction activity and illegal land deals. It is borne out by the 2011 census data which provide that there are 576,582 houses of which 125,503 are vacant. A large percentage of these may belong to the migrants (*First Post* dtd. 13.6.2013).

Spatial Distribution: There is broadly a spatial distribution of immigrants of different nationalities. For example, the Russian immigrants are largely concentrated in the northern parts of Goa such as the Morjim beach in Pernem *taluka*. Here, the immigrants have considerable influence on the services and hospitality sectors. Those from the west European countries have preferred the stretch in central Goa, between Vagator in the north and Benaulim in the south. The beaches further south, e.g. Palolem, Agonda, are largely preferred by people from the Middle East countries, particularly Israel. This zonal distribution patterns have at times created friction between the local population and the immigrants. This primarily results due to the differing lifestyles of the locals and the immigrants.

Consequences: The liberal, at times, permissive lifestyles of the westerners are at conflict with the traditional and rather conservative lifestyle of the village communities. There is an invasion by western culture; exploitation of traditional ceremonies, transfer of western values and behaviours, and an unintentional tendency to ape the western lifestyle. Proliferating resorts and spas are partly responsible for replacing and undermining the local culture. Socially engineered crimes are on the rise. The ethnically complex environment also serves as a refuge for criminals from other states (read as hostile countries). As such, there is the latent threat of potential catastrophe with the likely objective of settling international scores. Increase in narcotics, drugs and alcohol use/abuse, gambling (casino culture), human rights violations against women through the scourge of sex tourism, human trafficking and paedophilia are other social evils noticed in the tourism belt of Goa. The reported exploitation of school children by peddlers for drug trafficking (Herald dtd. 01.05.2023) is a matter of grave concern. There are also disturbing reports of women participation in drug trade. More serious is the question of not mere consumption and distribution of drugs, but the issue of manufacture of LSD (lysergic acid diethylamide-a potent psychedelic drug, known colloquially as 'acid') in the State of Goa. This is indeed a matter of grave consequence to the State and serious repercussions for the society (Times of India dtd. 03.05.2023). There is also the underlying challenge to the cultural fabric of the region which is getting modified and apprehensions are expressed in some quarters that perhaps it may get defaced.

7.8 Socio-Economic Constraints

The high demand placed on resources limits the options of distribution and sharing leading to confrontations and disputes.

7.8.1 Constraints in Sharing Resource

The immigrants and the large hotel conglomerates being financially affluent have purchased vast areas of coastal stretch, thereby causing tension between the locals and the immigrants in sharing resources. In most places, the lands have been barricaded, restricting access of the locals to common areas such as beaches, parks, *lakes*, religious structures, among others. Instances have been noticed where access to farmlands, water bodies, religious structures, recreational spots especially to the beaches in remote places of the State, has been blocked, thus causing conflict between the native stakeholders and the immigrants/corporates.

7.8.2 Friction Between Communities

Concentration of immigrants has attracted an influx of migrants from the rest of the country in search of employment opportunities. A stage has come whereby availability of any resource is greatly constrained causing difficulties for the local populace in the day-to-day functioning leading to friction between communities. For example, the living standard in the State has gone up considerably as compared to that in the neighbouring states, both due to immigrations and tourism. As a result, there is larger influx of migrant labour from adjoining states to meet the skilled and unskilled labour requirement of the various industries. The State is also confronted with demand and supply of perishable goods with the result that societal disturbances in neighbouring states affect the economics of the local market. The influx of migrant population is also responsible for social tensions within the State (frontline.thehindu.com).

Selective Employment: There is selective preference among the local youth for white-collar jobs, thus in an attempt to meet their ever-increasing requirements, the youth often resort to unscrupulous and illegal means to acquire wealth. Disproportionate affluence or disposable income in the younger segment of the society has influenced the enjoyment pattern and the lifestyle in general. Overall tourism has been responsible to encourage official corruption, wealth inequities and heedless competition. The moral fabric of the State is thus at stake.

7.9 Education vis-à-vis Employment

Subsidized Education: The State has provided great impetus to education by subsidizing both education-related infrastructure development and the human resource requirement. In addition to providing subsidies on fees, scholarships and freeships for the curricular courses, the State provides free computers, subsidized transport, mid-day meals and even books for school students. Teachers can avail of interest-free loans and study leave with pay for further education. In a population of 14.5 lakhs there is one State University with 23 post-graduate departments, 26 undergraduate colleges which include 10 professional colleges, over 96 higher secondary schools, more than 396 high schools, 434 middle schools and over 1190 primary schools. In addition to these, there is one private institute of technology, in addition to several autonomous institutes and centres of higher education of other universities. However, the trained human resource has to migrate to other parts of India in search of employment. Although the State is not averse to industries, it does not encourage major industrial establishments except for a selected few, under the garb of restricting pollution. As such employment generation is not commensurate with the trained human resource output. Major industries too are apathetic towards the State, primarily because of the constraints of land, labour and power availability as also at times, due to instant local opposition, some of which may be sponsored and stage managed. Tourism in the State provides employment to nearly 30% of the total workforce (Gokhale et al. 2014).

7.10 Shifting Occupation Pattern

The employment pattern has undergone a sea change, especially after liberation and more particularly with the encouragement to the tourism sector. Fishery and related activities provide employment trade and economic well-being to the local population. Moreover, fish being an important food source rich in protein it is desirable that fishing activity is carried out in a sustainable manner. The State on its part has put in considerable efforts to improve the fisheries business. Consequently, the marine fish production has increased from 24,600 t in 1965 to 87,984 t in 2013 (Anon 2014). Goa contributes nearly 2% to the total marine fish production of the country, contributing 3% to the State GDP and 17% to the agricultural GDP (Shreekant and Mujawar 2021).

7.10.1 Fisheries Sector

Fishing, which used to be a traditional occupation of the locals along the coastal belt, has been now overtaken largely by the migrant labour (Anon 2014). It provides livelihood to a large section of the society. Nearly 5–10% of the total population of the State is engaged in fishing and allied activities. The population associated with this industry is estimated to be 30,225 of which nearly 30–35% forms the active fisherman population as of 2013 (Anon 2014). The sea-food industry supports approximately 1.2 M jobs including those of harvesters, processors, dealers, wholesalers, retailers and the like. There are about 9,140 local fisherman and about 12,500 floating crew of the fishing vessels. Both the mechanized trawlers which number 843 (2013) and the motorized crafts which are about 780 in number, employ migrant crew in addition to the crew of non-motorized country crafts.

The maximum sustainable yield calculated from catch and effort data is of the order of 85,407 t (Anon 2014). In recent years, satellite-assisted fishing, based on the bio-optical parameters, is being increasingly employed to identify favourable locales of high value fish species (Shreekant et al. 2016). The annual fish catch although variable, in 2013 for instance, it was 92,662 t (Table 7.2) valued at ₹105,954 lakhs of which nearly 40% (value ₹33,939 lakhs) was exported (Anon 2014). Nearly 2/3 of the harvested fish is contributed by the mechanized sector. One of the traditional fishing gears used is the beach seine locally known as '***Rampon***'. The other popular gear the gillnet locally called '***Maag***' also plays an important role during the major part of the year. These fishing activities are labour intensive and hence generate employment and attract labour.

Brackish water fish farming, here referred to as Inland fish farming, is also an important activity in the region. The total area covered by the tidal flats (*khazans*) is approximately 18,000 ha of which about 3,800 ha is marshy and is suitable for aquaculture. In this sector about 600 ha is under pisciculture with sluice gate operations and about 175 ha is developed for shrimp farming. The freshwater bodies occupy

Table 7.2 Marine and Inland fish harvested in Goa during the period 2005–2013 (after Anon, 2015)

Year	2005 (t)	2006 (t)	2007 (t)	2008 (t)	2009 (t)	2010 (t)	2011 (t)	2012 (t)	2013 (t)
Marine	103,087	96,326	91,185	88,771	80,687	87,062	86,185	86,628	87,984
Inland	4194	4131	2447	3077	3283	3311	3538	3887	4678
Total	107,281	100,457	93,632	91,848	83,970	90,373	89,723	90,515	92,662

an area of nearly 3,300 ha covering Salaulim, Anjune reservoirs, Mayem *lake* and other smaller water bodies (Anon 2014). The State has taken initiative to develop freshwater aquaculture as it can help in improving local economy by providing extra earnings as well as generating employment. The nets, boats, motors and fuel are all subsidized by the State, yet surprisingly there is absolutely no price control on the fish catch when it reaches the open market, consequently the taxpayer has to bear the brunt silently.

Since the liberation of the State, however, the scenario in the fishing sector has changed considerably. The fishing community is diverting to hospitality sector, amusement and watersports whereas the farming community is getting attracted towards services and hospitality sector, and transportation. A section of both these communities is taking interest in sand extraction (mining) which is a lucrative occupation, considering the high demand for sand in the construction sector. Marine shell extraction for the manufacture of lime is yet another occupation of interest. The returns are high as the lime manufactured from shells is of better quality due to higher percentage of CaO and as such fetches better price.

7.10.2 Outmigration of Locals

The skilled and unskilled labour required by the infrastructure industry is met through the migrant influx. The literacy rate being high, the educated local youth are usually averse to the traditional occupations and prefer to migrate abroad particularly to the East African countries (prior to the 1970s), to middle east countries, not to mention the greener pastures in the western world, Canada and Australia in search of better employment opportunities. An estimated 55,000 locals are employed in the Gulf countries including Saudi Arabia, UAE (Dubai, Sharjah, Abu Dhabi and others), Oman, Bahrain among others (da Silva Gracias 2000). By and large the locals from south Goa, particularly Salcete *taluka*, have always preferred the hospitality sector, on board cargo vessels and cruise liners. As a result of this, farming (the traditional occupation of the locals in many parts of Goa) is generally avoided by the educated youth. Consequently, it is economically not viable for the owners to cultivate lands by employing labour. Thus, agriculture in the State has nearly collapsed. Despite the fact, large sums are being spent by the State to construct irrigation canals, most of which have never been used for the purpose for which they are meant.

7.10.3 Unviable Agriculture

A variety of agricultural subsidies such as free seeds, saplings at subsidized prices, fertilizers at concessional rates, reimbursement of costs for agriculture infrastructure and implements, crop insurance, support prices for crops and multiple other subsidies are provided by the State to make agriculture sustainable. Despite these efforts from the State, farmers have sold agriculturally productive lands to corporates. Barely 5% of the native population is engaged in agriculture. The total land under paddy cultivation, for instance, prior to liberation was 50,302 ha which in 2014–2015 has dwindled down to 41,970 ha. As per the 2011 census, there are 31,000 cultivators and 27,000 agricultural labour in a population of 1.4 M. Currently, nearly one-third of the State's 1.77 lakh ha of cultivable land is lying fallow (Anon 2015). These numbers suggest that agriculture is not an attractive proposition. Most people associated with agriculture are not fully dependent on it and have supplementary sources of income largely from tourism or employment abroad. It is uneconomical to till the land due to increased labour cost. In comparison, cashew cultivation which is fairly less labour-intensive activity, has picked up from 32,517 ha in 1960–1961 to 56,079 ha in 2014–2015 (Anon 2015). Thus, practically overnight, fertile lands and orchards became proposed construction/development sites. Large tracts of cultivable land are lying fallow only to be encroached upon by multinationals and builders for the development of infrastructure for commercial purposes. Thus, land prices have skyrocketed. There is also an increasing trend, to lease orchards and agricultural lands in the vicinity of the seashore, to the hospitality industry on a seasonal basis, as it is more lucrative than agriculture. The State has notified 112 proposals to convert over 5000 ha of prime agricultural land for settlement purposes (*Times of India* dtd. 6.7.2020) and has placed them in public domain for suggestions and objections. This also provides a measure of the aversion to agriculture in preference to construction or other activities.

7.11 Changing Lifestyle

Changing Living Standards: Social scientists are of the opinion that the native Goans do not prefer physical work and low-end jobs. However, it is also true that they do the same type of jobs as migrants to the Middle East or Europe. The current trend in the State is the locals have adapted to new living standards. International quest for vacation houses is forcing local residents out of their homes. During the tourist season, the residential premises are made available to international tourists as vacation houses and the owners manage in out-houses. Residential premises are also utilized to run seasonal hotels and restaurants patronized by low-budget tourists (Wilson 1998). Skilled and unskilled locals are employed in service industry, restaurants, hotels, transport companies, manufacturing companies for a part of the year and rarely practice their traditional occupation for the rest of the year. There is thus an acute shortage of local artisans and agricultural and farm labour.

Means of transport have undergone a change too; public transport, which used to be the only option available in rural area, is now the last preference. The stress is on the use of personalized vehicles. Bicycles are replaced by motor-cycles and the latter by cars. Bullock carts are replaced by motorized three wheelers or mini-trucks; ox-pulled ploughs are substituted by motorized manual tractors. Food habits have also undergone a transformation, fast food is replacing traditional cuisine. Weekly family outings for entertainment, recreation, including dinning out, are common.

7.11.1 Migration to Cities

Peer pressures are so tremendous that families are forced to shift their base from the villages to the cities as evident from Table 4.8, consequently affecting, housing, transportation, education, health care, recreation, entertainment and even religious rituals. It is estimated that 2.6 lakh Goans moved from their native place to another location within the State between 2001 and 2011 (*Times of India* dtd. 14.8.2019). The disposable income and availability of easy and rich money are not being wisely invested in productive ventures. Instead, it has brought in various societal evils such as free-spending frenzy on luxury 'dream' products and services, consumer lifestyles, flesh trade, gambling among others evils. This has in turn caused considerable stress to families and adverse influence on the juvenile local population. Greed and consumerism are on the rise and have led to devastation of families in many of the remote coastal villages from North Goa. The situation is quite alarming.

7.12 Concerns of Fragile Economy

Background: The development and prosperity of a region are primarily governed by its natural resource base and are sustained through thoughtful planning coupled with an efficient management. Both are commensurate with the rate of exploitation of the resources during which an inevitable impact on the environment is implicit. For integrated development, the fundamental objective is to strike a balance between conservation and development.

Fragility of Economy: Despite the various regional plans of the State, it has not been able to grapple with the intricacies of sustainability (Anon 2011). Exploitation of the resources and the trend of development have both been wayward/lopsided. In a small State, too much stress is laid on the mineral resource and tourism to generate revenue required to sustain developmental activities. Either of these resources fail to deliver, the State is pushed on the course of unrest/agitation and it is in a state of economic turmoil. This has been observed recently subsequent to the imposition of the mining ban. A similar discomfiture had occurred in 2008–2009 when the inflow of tourists had declined subsequent to a terror strike in Mumbai. This shows the fragility of the State economy.

It appears that too many initiatives have been placed on the development of tourism without assessing the carrying capacity of the State. This will have to be addressed from the lowermost rungs of the administrative pyramid such as creating and maintaining eco-villages, eco-municipalities, sustainable agriculture, green buildings, sustainable cities, green technologies, renewable energy and so on, without losing focus on the conservation of resources from land, water, air and of course, the humans. Although this is well said, it is difficult to achieve, however, a concerted effort has to be made to accomplish the objective.

Diversions to Release Stress: As mentioned earlier, the coastal belt is the one which is under tremendous stress. Tourists permeate every beautiful village along the coast, however, remote. Instead of leaving the entire coast open to tourism, there is a need to divert tourists from the beaches to selected localities in hinterland, thereby opening up new areas. This will serve to restrict free access to every place as well as help dovetailing leisure and recreation tourism with historical tourism, pilgrim tourism among others. This would enable relieve pressure on the excessively stressed coastal zone and consequently on the fragility of the economy. Sustainable tourism and ecotourism are two possible routes to contain the innumerable environmental and social problems. However, despite the increased popularity of the term 'sustainability', the possibility that the society will ever achieve sustainability is questionable, in view of excessive environmental degradation, overconsumption of resources, pursuit for infinite economic growth, climate change among others.

7.13 Increase in Illegal Land Conversions and Land Disputes

Inequitable Distribution of Resources: There is an increased awareness amongst the locals regarding the inequitable distribution of resources due to immigration and migration which the region is unable to cope up with. There has been an unprecedented statewide unrest that has compelled the State to take adequate steps to control large-scale land conversions and land transfers to multinational agencies. The Regional Plan of the State 2021 has been approved for implementation.

Prioritization of Activities: The decline in the environmental status of the coastal ecosystems as a result of non-sustainable use is an issue of grave concern. It is, therefore, essential to prioritize future activities in these areas for sustainable use. It is apparent that the State also lacks proper long-term planning. This is evident from the number of infrastructure projects completed within the last decade and few of which have been demolished or not being used within a few years of completion, to be replaced by new projects. This clearly shows that adequate thought process was not involved in the first place at the planning stage.

7.13.1 Policy Amendments

In order to alleviate further traffic congestion in cities and control overcrowding in the towns and villages, the State in consultation with the municipal councils, local self-governing bodies and other stakeholders should amend the policy on housing for cities, towns and villages, particularly in the coastal belt. This should outline the permissible limits of population density and, therefore, suggest restraint on infrastructure development (high-rise structures, housing complexes, hospital complexes, educational institutions among others) and other urbanization activities, such as commercial complexes. Similarly, construction of satellite townships and commercial habitats should be encouraged in the peripheral areas of major cities and towns.

All new infrastructure projects should mandatorily provide for vehicle parking areas within their own premises. Such measures will considerably reduce both traffic congestion and overcrowding which in turn will have a soothing effect on natural resources primarily land, air, water and consequently on human health and quality of life. This obviously will have a cascading effect on controlling air, water, land pollution and indirectly will enable to control problems related to waste generation, waste disposal, greenhouse gases generation, carbon footprints among others.

Land conflicts remain endemic. There are a total of 25,581-plus civil matters pending before the courts of law (*Times of India* dtd. 16.5.2022), of which, 2674 matters are pending for over a decade. There should be a robust statutory framework to settle land disputes. The establishment merely tinkers at the margin of the existing statutory provisions instead of bringing about fundamental reforms to rules and institutions. Far-reaching changes are desirable in the institutional structure which has largely remained static for several decades. All efforts should be aimed at good practices in managing coastal areas with due regard to conservation and optimum utilization by using minimum of all natural resources. In modern times resource scarcity is increasingly perceived as one of the greatest security risks of the twenty-first century.

7.14 Challenges to Development Initiatives

Background: After liberation (post-1961) of the State, there existed two strong political lobbies in Goa. One was pro-merger of the territory with the neighbouring state of Maharashtra and the other to maintain it as an independent entity. Considering the distinct nature of the society as well as of the problem, it was decided by the Central Government to assess the opinion of the people by conducting opinion poll on this issue. The anti-merger view prevailed and the territory was retained as a Union Territory within the Indian Union. In due course of time, the demand for statehood slowly evolved. One of the arguments then proposed in support of statehood was the small size of the territory. The size was projected as an asset for administrative purposes.

It was visualized that full potential of the State would be exploited for economic development and growth would be at a faster pace. Although it did turn out to be true and development did take place in several sectors including education, health care, infrastructure, sanitation, housing, electricity and water supply among others, it brought with it several shortcomings generally not witnessed in larger states of the country. It is only when the growth is affordable and inclusive can it be sustainable.

Although development has been at the forefront of the progressive agenda of the State, it is plagued with several constraints. The development agenda is often 'opposed as soon as proposed'. This has been observed in several instances of the State's developmental activity. For example, one of the severe constraints for development has been the opposition to any proposal brought forward by the government of the day. The opposition at times has been on flimsy grounds and ill-founded public opinion and this has been the nemesis of the State. Quite often the pretext of protecting the environment has been employed as a ploy for opposition and to obtain funding from abroad to sustain the resistance activity. The indices of development in which the State leads have themselves been instrumental for its pull-back too. For example, size, demography, education, infrastructure, socio-economic development among others have threatened the State and have held it at ransom. A few examples would be sufficient to illustrate the point.

7.14.1 Conflicts with the Industry

In the mid-eighties one of the multinational companies which also deals with paint manufacture intended to establish a plant to manufacture a synthetic product known as Nylon 66 at Keri in Ponda *taluka.* Although by definition any chemical plant is detrimental to the environment in general, with adequate safety measures chemical plants do function all over the world and are also required for the development and progress of the region. In the State, however, massive agitations were organized and finally the plant was withdrawn and shifted to Tamil Nadu where it has been functioning with a whole lot of safeguards for the last twenty-five years. Same was the case during the construction of a steel alloy plant called Metastrips. The cause of concern was the plant was to be located on the Verna *plateau*, and the effluents were to be released in the Zuari valley. In this case, the plant was approved with stiff safeguards. It remained functional for some time but subsequently was closed down due to other functional problems.

7.14.2 Waste Disposal

Systematic solid waste (garbage) disposal is one of the prime requirements of any developed society. During the last at least thirty years, several governments of the day have selected sites for disposal and treatment of garbage, but in vain. All attempts to establish garbage treatment facilities, with one recent exception, have been opposed tooth and nail, either by local self-government bodies or by the local population at large, with the result a socio-economically advanced State had no garbage disposal facility until recently. However, a small treatment plant has been now commissioned at Saligao. By the same logic, the fate of the proposed biohazardous waste treatment plant at Kundaim and the hazardous waste treatment plant at Dharbandora appears to be a distant dream.

7.14.3 Public Infrastructure

Public utility endeavours such as construction of railways, roads, waterways, airports are commensurate with the pace of development of a region and, therefore, are the need of the hour. Construction of N-S coastal railway line was proposed by the Central Government and it was due to pass through the State on its way to Karnataka and Kerala in the south. While most riparian states welcomed the decision, it was objected to in Goa. Initially, the agitation was to seek a realignment of the rail route through the hinterland in order to protect the *khazan* lands and to improve the means of communication of the relatively underdeveloped region of the State. However, once that demand was turned down, further the agitation continued on minor issues such as construction of subways, foot overbridges and issues such as spread of malaria due to migrant labour. Finally, the matter had to be resolved in the court of law and thus a national project was delayed beyond its schedule. It must be noted, however, that the damage to the *khazan* lands and the Carambolim *lake* ecosystem has been quite severe. This could and should have been prevented.

The same holds good as regards the construction of the airport at Mopa, the laying of a railway track parallel to the existing South-Western Railway line through Goa, four/six lane state/national highways and so on. More than taking advantage of the development and the employment-generating capacity of the project, what is projected for opposing the airport, for instance, is, the partial diversification of air traffic from the existing airport to the new one and, therefore, the purported loss of employment to the local populace in the vicinity of the existing airport.

In yet another instance, under the national project for improvement of infrastructure, grants for the construction of a four-lane highway were allocated to the State. The state highway would connect to the neighbouring states of Maharashtra and Karnataka leading to better connectivity and opening up avenues for increased trade and commerce. Whereas the highways in both the adjoining states are on way to completion and partly functional, the portion within Goa is now being constructed after considerable delay in obtaining clearance for the alignment, land acquisition along the alignment route and complete other environmental clearances. Loss of green cover is completely justified to press for a realignment of rail/road as the case may be, wherever possible and should be entertained by the State to the extent possible.

7.14.4 Resource Allocation

Sharing of natural resources with neighbouring states has been a problem with every state in the country and Goa is not any different in this regard. With growing demographic (tourism included) pressures and thrust to the industrialization coupled with world climate variability, the demand for water is bound to increase. Consequently, there is likely to be a spurt in interstate water disputes in years to come. The State is already confronted with a couple of them.

A few low order tributaries of Madhei (major tributary of Mandovi), at the source region in the Western Ghats, originate either in the states of Maharashtra or Karnataka. The Karnataka state proposes to divert one stream, the Kalasa-Banduri *nallah* to harness its waters. Similarly, Maharashtra also intends to divert waters from another stream, namely Kattika *nallah*, a tributary of Valvanti (minor tributary of Mandovi), to the adjoining basin of Tillari river, to augment the live storage. Diversions of natural water courses should not be encouraged in principle, since they impact the ecology and hydrogeology of the region in question. However, considering the increasing demand for water, diversions are being undertaken.

Though no exhaustive studies are available, the total yield of Mhadei basin (watershed area 899.46 sq km, 36% is in Karnataka) in all the three states is estimated by the eco-development plan task force (2014–2015) of the Planning Commission to be 4555 Mm^3 (as against the total yield of Mandovi basin of 9102 Mm^3). The quantity of water to be diverted from Kalasa-Banduri is 5.6 Mm^3 and that from Kattika *nallah* is 14.138 Mm^3 which is nearly 2.52 times that of former. The total quantum of water likely to be diverted is 0.43% of Mhadei yield and 0.21% of the total yield of the Mandovi basin. The impact of these diversions on the ecology and hydrogeology downstream needs to be assessed and is a matter of debate. Yet an impression has been created in the Government circles that the diversion is likely to adversely affect the hydrological and ecological balance of Mandovi basin and would therefore be unfavourable for the State. The Mhadei Water Dispute Tribunal has adjudicated over the matter of water allocation and the State has gone in appeal to the Supreme Court of India.

It must be noted that more than 40% of the waters from Mandovi, Zuari and other rivers drain into the Arabian sea, however, there are serious objections if a small fraction is harnessed by neighbouring states. A precious resource, such as water, has to be conserved on all scales and put to optimum use for the betterment of humanity irrespective of ethnicity, colour, cast, creed or even nationality, of course with the proviso, that it should not adversely affect the region downstream. It would be prudent for the concerned states to settle the dispute amicably in the interest of people of all three states with the help of a multidisciplinary expert panel.

It is pertinent to mention here that natural diversions have been taking place in geological past. For example, the picturesque Dudhsagar (waterfall) River which used to be an east-flowing tributary of Malaprabha, through Karnataka, in the geological past, is now flowing westerly as a tributary of Mandovi through Goa, because of river-piracy that occurred due to higher rate of headward erosion of Mandovi (similar is the case of Jog Fall or Gerusoppa waterfall on Sharavathi river). Thus, a cautious approach needs to be adopted in all such matters, lest diversion of tributaries of Dudhsagar may adversely affect the State on several counts. There have been instances in the past when Goa and Maharashtra states have cooperated with each other in sharing the waters of the Tillari river, a tributary of Chapora, for mutual benefit. The State has not been able to fully utilize the benefits of this joint venture whereas Maharashtra has reaped the benefits. Sadly, it merely reflects on the State's failure to do so. A wise counsel should prevail in such matters.

National Water Policy formulated in 1987 and later amended in 2002 and 2012 stipulates water to be under 'state jurisdiction'. Interstate disputes, however, are referred to tribunals established under Interstate River Water Dispute Act 1956 amended in 2012. Time is ripe that water should be treated as a 'national resource' and a policy formulated on sharing of water resources. All related issues such as quantum of water to be shared by each state, permissible extent of diversion and all related matters could be then decided by a multidisciplinary expert panel which shall arrive at decisions on the basis of detailed studies on ecology, hydrology, socio-economic and socio-cultural impacts on the regions in question.

Education: The establishment of a State University was proposed in the seventies but could start functioning only in the eighties after considerable opposition primarily related to site selection. Later, it became apparent that funding the university was a major constraint. Considering the financial crunch faced by the State and consequently by the University, a proposal was made to the Central Government to identify it as a Central University so that the funding could be borne by the Centre. The Central Government agreed in principle to grant a national status and a bill was duly introduced in the parliament to amend the existing provision, however, opposition from the very members of parliament representing the State led to withdrawal of the proposal. An excellent opportunity was thus squandered and the State lost a good prospect which could have provided avenues for furtherance of education and research in the State. National Institute of Technology allocated to the State is finally coming up at Cuncolim after a gap of more than eight years since the sanction. On this count, the establishment of the proposed Indian Institute of Technology, sanctioned to the State, appears to be a distant dream.

Other Initiatives and Constraints: The proposed Information Technology habitat at Taleigao *plateau* met with similar future. After acquisition of land and preliminary infrastructure creation, opposition to abolish the park started gaining momentum and ultimately the project was shelved. A Special Economic Zone (SEZ) was proposed for establishment at Balli, Salcete *taluka*. Issues connected with acquisition of land put the project in jeopardy. Projects of this nature draw special attention from the decision-makers as they are viewed more as construction projects.

Development projects of various kinds such as construction of housing complexes, hotels, food-parks, marinas, irrigation projects are all being opposed on various grounds, primarily (supposedly) to protect the environment and to control the onslaught of immigration and migration. Society's response should best be mature and focused on real issues. What is worse, of late, religion is often found to play an active role in such activities which are detrimental to the society in the long run. Such a mix-up of religion and politics can snowball into a potentially explosive concoction as has been observed globally from Croatia, Serbia to Iraq, Somalia, Afghanistan, Pakistan and back home in Kashmir. A secular and pluralistic society cannot afford this distraction.

The other major constraints are the availability of land and change in the land-use pattern. Land is one resource that has been under extreme stress with the result any initiative for developmental activity is hindered, due to land availability, consequently nefarious land deals and litigations are on the rise. The pressure on land is so acute that there is clamour already that even the key establishment, such as of defence, located at Panaji, since colonial times, should vacate the land and return it to the Corporation of City of Panaji (*Times of India* dtd. 14.11.2014) obviously for purposes other than defence.

It is clear by now that few human activities including mining operations or infrastructure developmental projects are without risk and environmental damage. Thus, what is required is an informed public to help understand and balance the risk of a potential operation with its rewards, since it is almost impossible to continue with developmental activities without impacting the environment.

7.15 Future Perspective and Recommendations

This study persuades us to think harder on the twin aspects of the conservation of natural resources and the conservation-cum-protection of the environment by efficient use of resources by way of recycling, substitutions and innovations, so as to maintain sustainable levels of all activities leading to the development and welfare of the local populace and the tourists. There is a considerable level of local discomfort with regard to the encroachment upon the beaches by land mafia. The sanctity of beaches as the most vital resource for tourism in the State needs to be maintained at all costs. The degradation as well as the pollution of all resources has to be promptly arrested and efficient waste management has to be ensured if the needs of future generations are to be sustained and met. The need to make the tourism industry sustainable by adopting appropriate rules and regulations is paramount. It would curb unscrupulous overexploitation of the precious natural resources of the State.

Need for Public Infrastructure: The upgradation of public infrastructure including roads, electricity, water supply, sanitation, waste disposal among others should get top priority. Simultaneously, there is great demand as well as scope, to ensure free and unhindered access to remote destinations of tourist-interest and to the village commons, the beaches, lakes, places of worship and the like.

Sewage System and Sanitation Economy: There is dire need for improvement and upgradation of the sewage system commensurate with the contemporary civic practices and facilities whereby the townships and semi-urban centres have an underground sewerage system with a sewage treatment facility. A wastewater recycling plant is an urgent necessity to ease sewage treatment, improve water quality and environmental upkeep. In the coastal areas, where underground sewage network is untenable, the construction of septic tanks is essential to discontinue with the use of pit-latrines completely. Efforts need to be directed to improve sanitation by joining forces with the sanitation agency of the State/private sector to enable new models of cost-efficient and revenue-generating sanitation system beneficial to both the State and private sector, respectively. This will give much needed boost to the tourism activity in the State providing greater incentives to all stakeholders including the tourists and potential service providers. It will contribute to sustainable tourism, facilitating the achievement of the United Nations Sustainable Development Goal (SDG)-6.

Sustainable Resource Management: The State should put in place a mechanism of continuous assessment of natural resources to ascertain the levels of sustainability from time to time and suggest mid-course corrections as and when required. Among these, water is one of the prime resources which indirectly influences the economy, is essential to enhance the quality of life of the people and hence needs to be attended to on a priority basis. Efficient utilization of the natural resources in a sustainable manner should be the focus of economic welfare and quality of life of any society. Any departures from efficient utilization by way of under-exploitation or overexploitation of the resources may result in long-term damage to the society, too difficult for redemption. Water resource management primarily involves three sectors among others, agriculture, generally the thirstiest consumer, accounting for 70% withdrawal, but not so much of a concern to this State, industry and energy with 20% demand and 10% for the municipal water supply.

Agriculture in small landholdings in the State is not viable under present economic conditions. Although theoretically the State needs to increase food production by more than 60% and possibly by 100% by 2050, on this count the State/sector is not sustainable. More than 50% of the resident population will be living in urban centres (Table 4.8) by 2050. The State is ill equipped for this rapid change and will have to adopt appropriate measures soon, if it has to contain the development of slums, already surfacing rapidly in major towns.

Although there is considerable discussion on sustainable development, the global efforts in that direction have not been completely successful. Environmental degradation, although not as worse as in many other places within the country and in other emerging economies worldwide, does not paint a very rosy picture in the State either. The sooner the State realizes the limitations of the boundaries of the life support systems, better it is. This will form the basis for future sustainable development framework.

Regulation of Resource Use: The quantum and rate of exploitation of minerals, water and all other resources to sustainable levels should be fixed from time to time, considering the resource availability/generation capacity, and the demand. The existing resources have to be shared and managed efficiently in economically gainful manner to accrue long-term benefits, not only for the present, but also for the future generation too. Blending, beneficiation (in case of mineral resources), recycling, substitutes, likewise various types of recharge, harvesting and recycling techniques need to be adopted to conserve resources and thereby maintain environmental safety and security. Special emphasis needs to be placed on, not only to efficiently manage the water resources of the State but also to collect, treat and reuse used water. In course of time, a separate drainage and sewage system may have to be put in place to facilitate reclamation of used water.

All mineral resources are scarce and hence are known as wasting assets. Therefore, efforts should be directed towards minimum use, recycling and reuse of used materials such as the construction and demolition waste, instead of overstressing the existing resources and concentrating merely on landfill which ultimately results in the contamination and degradation of yet another precious, life sustaining resource, namely water. In order to control pollution of soils and groundwater from industrial effluents the State should consider in due earnest of bringing in a regulation modelled on the 'Polluter Pays Principle' in addition to educating the erring companies to contain and prevent pollution before it occurs. The reduction in the use of nonrenewable material wherever possible could be yet another step in that direction.

Participatory Approach: As the economy of the region primarily revolves around tourism, the State has to develop partnerships with national and international organizations such as WTO, UNWTO, WTTC and other international tourist organizations in order to enlarge tourism to the next level and thereby augment the competitiveness of the local stakeholder such as the Goa Tourism Development Corporation, Goa Hotel Owners Association, Goa Shack Owners Association, Goa Taxi Union and others. The different type of enterprises operate on different scales and appears to help meet the objectives of the tourism planners, however, closer scrutiny shows that industrially scaled enterprises generally do not satisfactorily meet the objectives of economic planners, such as increased earnings, foreign exchange investment, employment opportunity, production, entrepreneurship, infrastructure and the minimization of adverse social and cultural effects. An innovative approach is the need of the hour to ensure that the economic benefits of tourism percolate to the local community rather than being soaked up by business operators and corporates promoting hotel and restaurant chains. The sustainability models accepted in principle have to be implemented on ground in letter and spirit.

Policy-Framework: The State tourism policy should focus on the deprived and marginalized sections of the society as defined by United Nations Commission for Sustainable Development, so that the fruits of tourism development benefit the lowermost rungs of the society helping their uplift. There exists a distinct development gap between the western coastal belt of the State where there is better infrastructure and the service sector is better established than the eastern hinterland region. It is necessary to bring about a multidimensional change, evenly spread over the entire State, rather than be restricted to a specific region, so much so that the benefits of the tourism sector are instrumental in changing the regional disparities. The sustainable development should avoid large differences in the distribution of resources and economic gains especially apparent between the coastal belt and the hinterland, so as to avoid social conflicts. There should be greater local participation in monitoring and implementation of environmental management plans. Large-scale migration to the State is a major concern, unless it is contained, the gains of sustainable development will not be appreciated and truly realized. A sustainable migration policy will have to be devised to attain sustainable development.

The future demands the formulation of clear and inclusive legal policy with simplified procedural requirements oriented towards balanced, long-term development with sustainable growth where knee-jerk developmental projects with short-term gains and sometimes disastrous/harmful consequences have no place.

Diversification: Tourism activity should be regulated to sustainable levels by avoiding mass tourism. The mass-tourism arrivals threaten the environment in all seriousness, particularly the beaches; it is ***the most essential resource on which the tourism has been developed***. More efforts should be directed towards diversification and further development of other types of tourism such as ecotourism, cruise tourism, archaeology tourism, heritage tourism, cultural tourism, medical tourism, wellness tourism, wildlife tourism, film tourism, adventure tourism among others to relieve the beaches from excessive stress. More activities with a focus on the beaches should be avoided, to relieve the beaches from added strain. Considering the constraints on land and water resources, there is a need to avoid tourism products that are resource intensive. Similarly, the State should be vigilant enough to curb tourism that jeopardizes the safety of women and children, leading to increased crime and antisocial activities including drug peddling, substance misuse, gambling, prostitution and paedophilia.

Encouragement to Small Enterprises: There is a need to project environmental conservation as a prerequisite to tourism. National and local stakeholders have to be encouraged to develop small enterprises based on conservation of nature and preservation of environment.

Tourism Audit: Periodic environmental audit, including the social and cultural dimensions of tourism, is a must to determine the carrying capacity of the natural resource to guarantee the preservation of the environment and ensure safety of the people and the ecosystem. A transparent, participatory database on the resources of the State is long overdue.

Developing Linkages: The State machinery should join hands with the private sector, the local self-governing bodies and the local community to develop strategies for linking tourism with other sectors of economy, society and the environment involving various associations and organizations that deal with tourism-related activities. The tourism activity should be undertaken as a teamwork with better intra- and inter-community linkages and cooperations.

Overcoming Competition: The tourism industry in Goa faces a stiff challenge from neighbouring states and other emerging economies in the east such as Malaysia, Thailand, Cambodia, Vietnam to name just a few, where varied and innovative tourism portfolios are marketed at competitive rates, at times much less expensive, to attract international tourists. The stakeholders in Goa have to remain vigilant to these trends in this field. The maintenance of cleanliness of the beaches and the environment in general, infusion of a courteous hospitality with prompt and disciplined service, particularly to the domestic tourists, by eschewing unethical practices at all levels, including at the level of agencies maintaining security, law and order, is a primary requisite.

Stimulation of Community-based Tourism: The involvement of local community stakeholders such as local artisans and craftsmen in the tourism activity and extending it to the involvement of agriculture sector would go a long way not only in making it more appealing to the tourists but would support the community, boost economy, protect the environment and make it more sustainable.

Empowerment of Women and the Underprivileged Sections: Considering the resource constraints, the level of environmental degradation, massive waste generation and the constraints of efficient waste disposal, the State should ensure that a large majority of stakeholders actively participate in this sector and get benefitted from it. This can happen when the disadvantaged, underprivileged and socially marginalized sections of the society and women, in particular, to minimize gender discrimination, actively get involved. If the benefits of tourism need to percolate to these sections, efforts need to be directed and accelerated towards their involvement in the tourism activity coupled with the State-envisioned tourism, by encouraging small community-level enterprises to participate in the tourism activities. Often the economic benefits promote voluntary involvement. Apart from environmental protection, both economic growth and social inclusion of this section of the society are crucial to harmonize sustainable development.

Moreover, it has been globally recognized that low-budget tourism is the least destructive and most beneficial path that will involve maximum stakeholders and provide them a means of livelihood, in the nation's goal of achieving 'Aatma Nirbhar Bharat' (self-reliant India), at the same time it would help conserve resources in a sustainable manner. If that is achieved it will be the greatest service to the people of this State.

Capacity Building: Tourism industry being labour intensive and multifarious in activity requires human power of varied skills and in great numbers. Therefore, there is a need for a continuous exchange of views and ideas between the public and the private sector to understand each other and to develop a roadmap of sustainable manpower development in Goa. This should encourage institutional capacity building in tourism and other sectors such as environment and culture through awareness and training programmes to facilitate the use of modern technology, communication and information system in the process of achieving sustainable tourism development.

Academic Initiatives: Encouraging studies related to Goa-specific tourism and its potential linkages with other forms of tourism within the country and internationally could be introduced in the educational framework of the State. Similarly, a component of sustainable waste management including solid waste, biomedical waste, construction and demolition waste should be included in the educational curricula at various levels and different disciplines to bring about an awareness and sensitize young minds on protecting the environment. Such initiatives may eventually develop into establishing an Institute of Tourism Studies with specializations or general focus. The primary initiative of such programmes would be to pursue tourism growth and remain competitive, especially with mature tourist destinations and the emerging ones. Such academic initiatives would promote public institutions and tourism administration in terms of WTO to stimulate tourism growth in a sustainable manner, by attracting international tourism, formulating strategies for coexistence of tourists and local communities and minimizing unruly and disrespectful behaviour, and facilitating the much required innovation in improving the knowledge base of this recent phenomenon.

Awareness and sensitization programmes and workshops need to be organized for the civil engineers, architects and contractors from the construction sectors regarding the innovations on various aspects of resource management, use of substitutes, recycling of materials, innovations in product substitutes and the like. Likewise, awareness/training programmes and workshops for local civic bodies such as panchayats, municipalities, government departments and industry and people at large could be organized towards capacity building in production of sustainable materials.

These developments would initiate a thought process whereby innovative ideas would generate from a bottom-up participatory approach between the stakeholders and the sector in concern, so vital for any development. Sharing information and knowledge between and among the State machinery, the civil society and experts from research and development organizations would go a long way in building on the experiences and developing innovations in attaining the objective. Philanthropic/international donors may be willing to assist in such initiatives which are beneficial to the community at large.

Fostering a Low Carbon Footprint: In the present global scenario, all efforts should be directed towards evolving a low-carbon footprint and thereby contribute to the efforts of the world community to controlling global warming by maintaining it below 1.5–2.0° C. The State should motivate various enterprises to actively participate in Green Revolution. Efficient use of public transport will be a step in the right direction to reduce fossil fuel consumption. It should incentivize on the generation of clean and green energy by contributing to the nation's 167 GWh power generation from renewable sources. It should provide better healthcare facilities that cater to all sections of the society. The key to the well-being and healthy cities/societies lies in urban planning, well-organized public transportation, environmental regulation and providing healthy food. The answers to sustainable development lie at the intersection of public health, nutrition, urban planning, transportation, education and law enforcement. There should be a concerted effort to disincentivise development that destroys the building blocks of our health. The State needs to invest more on the generation of renewable energy resources, conservation and rejuvenation of natural resources, with a strong and vigilant executive body armed with requisite regulatory and enforcemental powers. Multi-disciplinary and multi-sectoral collaborations need to be encouraged. It may be reiterated that it is only when the growth is affordable, inclusive and founded on renewable resources can it be sustainable. Finally, all out efforts need to be directed towards reducing the carbon footprint at all costs. Ensuring a smooth transition from a fossil fuel-based economy to a renewable one is the inevitable solution in proceeding towards climate neutrality.

References

Anon (2011) Goa development report. Planning Commission, Government of India, New Delhi, Academic Foundation, New Delhi, 1–250

Anon (2014) Goan fish trails. An overview of Department of Fisheries, Government of Goa, 1: 1–40

Anon (2015) A report on agriculture input survey 2011–12 of Goa state. Directorate of Planning Statistics and Evaluation, Government of Goa, 1–388

Anon (2021) Official Gazette. Government of Goa, Series 1, No. 23, 1213–1244

Brundtland Report (1987) Our common future. United Nations. http://www.are.admin.ch

CRZ Coastal Regulatory Zone (CRZ) Notification (2011) Ministry of Environment and Forests, Gazette of India, Extraordinary, Part-II, Section 3 Sub-section (ii) of 6th January, 2011. http://www.environmentalportal.org.in

CRZ Notification (1991) Notification under Section 3 (1) and Section 3(2)(v) of the Environment Protection Act, 1986 and Rule 5(3)(d) of Environment (Protection) Rules, 1986, Declaring Coastal Zone (CRZ) and Regulating Activities in the CRZ, Gazette of India Extraordinary, Part II, Section 3 Sub-section (ii) of 20th February, 1991. http://www.environmentclearance.nic.in

Da Silva GF (2000) Goans away from Goa. Lusotopie 2000:423–432

Goa Tourism Policy (2020) http://www.goa.gov.in

Gokhale KD, Sawant N, Ugavekar N (2014) Goa: Tourism and sociocultural implications a study of selected destinations of South Goa district, Goa-India. IOSR-JHSS 19:36–41

Hall CM (1998) Tourism and politics, policy, power and place. John Wiley and Son Chichester. http://www.sciepub.com

Hall CM, Jenkins JM (1995) Tourism and public policy. Routledge, London and New York, p 116. http://www.worldcat.org

Hickel J (2020) The sustainable development index: Measuring the ecological efficiency of human development in the Anthropocene. Ecological Economics 167. sustainabledevelopmentindex.org

Hillali M (2007) La politique du tourisme au Maroc- Diagnostic, bilan and critique. L'Harmattan, Paris, p 296. http://www.decitre.fr

Jagtap TG (1986) Ecological studies in relation to the mangrove environment along the Goa coast, India. Ph. D. thesis, Shivaji University, Kolhapur (Unpublished)

Jenkins CL (1980) Tourism policies in developing countries: a critique. Int J Tour Manag, March, 22–29

Lamb AN (1998) Tourism development and planning in Australia- the need for a national strategy. Hosp Manag 7:353–361

Lea J (1988) Tourism and development in the third world. Routledge, London, 14, pp 1–88. https://doi.org/10.1177/030913259001400113

Menon KAG (1993) Case study on the effects of Culture and Environment India, Jaisalmer, Khajuraho and Goa. UNESCO, Principal Regional Office for Asia and the Pacific, Bangkok, pp 1–114.

Sawkar K, Noronha A, Mascarenhas A, Chauhan O (1998) Tourism and environment: issues of concern in the coastal zone of Goa. The Economic Development Institute of the World Bank, pp 9–18. www.drs.nio.org

Shreekant GB, Subramanian S, Lekshmi M, Singh NP (2016) Validation on the satellite based potential fishing zone advisories along Goa, SW coast of India. Indian Journal of Fisheries 63:8–15

Shreekant GB, Mujawar S (2021) Fisheries sector of Goa: present status, challenges and opportunities. National Seminar on Life and life processes: integrated approach for sustainable development, Goa University virtual mode. www.researchgate.net

Trinder M (2020) Cuba found to be the most sustainably developed country in the world, new research finds. Studio for Sustainability and Social Actions, Article 2. sites.psu.edu

UN General Assembly (2005) Environment and sustainable development. World Summit 14–16 September, New York. un.org

Untawale AG, Dwivedi SN, Singbal SYS (1973) Ecology of mangroves in the Mandovi and Zuari estuaries and interconnecting Cumbarjua Canal of Goa. Indian Jour Mar Sci 2:47–53

Velasco M (2018) Tourism policy. Global encyclopaedia of public policy and governance. Springer International Publishing AG, pp 1–11

Venugopal T (2021) Tourism and sustainability in India- exploring sustainability of Goa tourism from the perspective of the local community. Eur J Manag Bus Econ 6:34–41
Wilson D (1998) Paradoxes of tourism in Goa. Annals of Tourism Res 24:52–75
WTO (1997) Towards new forms of public-private sector partnership: the changing role,structure and activities of national tourism administration. World Tourism Organization, Madrid. wto.org
Zappino V (2005) Caribbean tourism and development: an overview. Discussion paper No. 65, European Centre for Development Policy Management, pp 1–38

Index

A
Acacia auriculiformis, 154
Afforestation, 55, 68, 149
Aggradation, 160
Agonda, 10, 31, 208
Agricultural subsidies, 213
Aguada, 9, 10, 22, 23, 31
Air, 11, 41, 44, 46, 68, 76, 129, 141, 143, 144, 146, 147, 149, 150, 152, 155, 162
Ambaul, 10, 198
Anastomosing, 8
Anmode Ghat Trondhjemite Gneiss, 20
Anmode palaeoplain, 30
Anthropogenic activities, 98, 118, 119, 121
Aquatic life, 67, 124, 128
Aquifers, 52, 73–77, 81, 82, 84, 90, 92, 93, 96–100, 106, 108–111, 120, 124, 133–139, 141, 142, 159, 177
Archaean, 16, 18, 20, 22
Argillite, 20, 21
Arid, 72
Artificial recharge, 109, 110
Arvalem, 24, 94, 120
as Bacteria, 148
Atmosphere, 29, 41–44, 47, 73, 106, 131, 143, 144, 149, 178

B
Bababudan, 18
Bababudan Group, 19
Bacteria, 127, 130, 133, 141
Baga, 7, 10, 79, 81, 136, 137, 202
Banded Hematite Quartzite (BHQ), 21, 23–26, 94
Banded Iron Formation (BIF), 20, 24, 25
Banded Magnetite Quartzite (BMQ), 26
Barazana, 25
Barbed, 8
Barcem Formation, 19, 20
Barcem Group, 19, 20
Basement problem, 9, 17–19, 82
Bati, 120
Bauxite, 8, 30–32, 57, 59, 61, 179
Beaches, 2, 10, 34, 35, 130, 166, 180, 199–201, 208, 209, 215, 221, 224
Beach sand, 34, 81–83, 166
Belgavi, 194
Benaulim, 10, 11, 208
Beneficiation, 52–54, 111, 122, 123, 150, 222
Bicholim, 11, 20, 23, 25, 84, 85, 94, 96, 120, 145, 175
Biochemical Oxygen Demand (BOD), 125, 126, 134, 140, 141
Biodegradable waste, 175, 197
Biodiversity, 11, 41, 116, 120, 121, 124, 127, 190, 191, 193, 201
Biogas, 143, 174, 175
Biohazardous waste, 218
Biomass, 46, 47, 55, 101
Biomedical waste, 175, 176, 225
Biomethanation, 175
Biosphere, 41, 43, 44, 101
Biota, 116, 121, 122, 126, 127, 131, 136, 159–162, 180
Bondla, 19, 20, 22
Bottom ash, 174

A. G. Dessai, *Environment, Resources and Sustainable Tourism*,
Advances in Geographical and Environmental Sciences,
https://doi.org/10.1007/978-981-99-1843-0

C

Calangute, 134
Calcination, 35, 178
Canacona, 5, 19, 20, 22, 23, 31, 33, 85, 96, 105, 120, 134, 136, 201
Cananguinim, 9
Cantor, 120
Carambolim, 11, 120, 218
Carapace, 31, 35, 157
Carbon, 11, 27, 33, 40, 42, 43, 144, 156, 174, 182, 191, 197, 216, 226
Cardice, 126
Carrying capacity, 27, 46, 48, 126, 128, 155, 164, 194, 195, 215, 224
Castlerock Band, 5, 18
Casuarina alstonia, 154
Chandranath Granite, 19, 20
Chapora, 7, 9, 33, 34, 36, 79, 81, 105, 136, 158, 220
Chauri, 5
Chemical Oxygen Demand (COD), 125, 134, 135, 140, 141
Chinchinim, 120
Chitradurga Group, 18, 24
Chitradurga Shear Zone, 16
Chloritic division, 5
Chopdem, 36, 164
Chorao, 120
Chorla ghat, 7
Climate, 4, 11, 32, 35, 41, 72, 73, 101, 143, 149, 171, 172, 191, 215, 219, 226
Closepet Granite, 16
Coastal aquifers, 97–100, 106, 133, 134, 142
Coastal regions, 41, 49, 98, 100, 193
Cola, 10, 198, 201
Coliform, 125, 133–135, 140, 141
Colva, 10
Concrete waste, 177
Conjunctive use, 108, 111, 112
Consequent, 7, 120
Conservation, 5, 23, 36, 40, 44, 51–54, 59, 68, 73, 74, 79, 93, 104, 105, 107–109, 111, 112, 116, 126, 137, 143, 163, 165, 172, 180, 188, 193, 214–216, 221, 224, 226
Construction and demolition waste (CDW), 171, 172, 176, 177, 208, 223, 225
Consumerism, 214
Corrosion, 177, 178
Cortalim, 31, 120, 121
Cost–Benefit Analysis (CBA), 149, 150, 152
Cumbarjua Canal, 8, 35, 121, 124, 201
Cuncolim, 135, 136, 140, 175, 183, 220

D

Dando, 95
Deccan Traps, 5, 7, 16, 18, 20, 22
Demolition waste, 171, 176, 208, 223, 225
Dendritic, 8
Development, 4, 5, 9, 23, 27, 30–33, 41, 44–46, 48, 50, 53, 54, 56, 58, 61, 62, 64, 81, 83–85, 92, 93, 95, 97, 103–105, 107, 111, 112, 116, 117, 121, 122, 137, 140, 142, 147–150, 152, 154, 165, 177, 181, 186, 188–197, 199, 203, 205, 206, 210, 213–218, 220–226
Dharbandora, 85, 183, 218
Dharvel, 10
Dharwar Supergroup, 24
Dharwar System, 5, 18
Disposable income, 210, 214
Distributary, 8, 121
Dolerites, 20, 22, 23
Dona Paula, 9, 10, 31
Dry ice, 126
Dudhsagar, 5, 8, 11, 219, 220
Dunes, 34, 82, 139, 203
Duricrust, 29, 31, 81, 82, 94, 95, 157
Dust pollution, 117, 150
Dyke swarm, 22, 23

E

Earth, 4, 40, 44, 47, 51, 58, 59, 72, 73, 116, 147, 176, 191
earth, 40–43, 46, 47, 49, 51, 73, 79, 116, 143, 148, 153, 155, 165, 166, 188, 201
Eastern Dharwar Craton (EDC), 16
Ecology, 4, 5, 44, 45, 54, 55, 64, 103, 155, 159, 160, 182, 188, 189, 195, 197, 201, 219, 220
Economy, 2, 4, 10, 23, 27, 40, 50, 51, 58, 72, 107, 112, 118, 127, 154, 155, 172, 179, 182, 188, 189, 191, 193, 195, 197, 199, 200, 206, 207, 212, 215, 222–226
Ecosystem, 34, 40, 49, 102, 121, 125–131, 141, 143, 150, 160, 163, 165, 191, 198, 201, 203, 215, 218, 224
Ecotourism, 199, 215, 224
Education, 45, 54, 118, 148, 165, 184, 190–192, 210, 214, 217, 220, 226

Effluents, 102, 121, 122, 126, 128, 135, 142, 143, 150, 174, 175, 183, 184, 217, 223
Electronic waste, 179
Emission, 26, 29, 31, 32, 46, 107, 144, 147, 149, 150, 178, 184, 191, 192, 200, 205
Employment, 4, 54, 58, 142, 157, 163, 164, 179, 183, 184, 188, 189, 191, 193–195, 197, 199, 200, 209–213, 218, 223
Environment, 4, 5, 8, 16, 18–20, 23, 26, 27, 40–42, 44–46, 48–51, 53, 54, 56, 60, 64, 67, 93, 102, 103, 110, 116, 117, 120, 122–131, 134–136, 141, 143, 147, 149–152, 154, 155, 157–159, 161, 164, 165, 171, 172, 176, 179, 181–184, 188–192, 195–198, 209, 214, 217, 220, 221, 224, 225
Environmental degradation, 116, 142, 165, 215, 222, 225
Environmental management, 53, 55, 149, 223
Escarpment, 6, 8, 30, 64
Estuaries, 8, 34–36, 96, 99, 120, 126, 139, 158–161, 165, 198, 200
Eutrophication, 184

F

Facies, 5, 20–22, 24
Fauna, 123, 127, 155, 163, 165
Ferricretes, 30
Flesh trade, 197, 214
Flora, 123, 127, 155, 165, 201
Flux, 42, 126
Fly ash, 148, 174, 177, 178, 181, 208
Flysch deposits, 21

G

Gabbros, 22, 34
Galgibaga, 7, 10, 79
Garbage, 67, 131, 133, 141, 143, 172, 173, 175, 180, 217, 218
Gases, 26, 31, 41, 46, 132, 143–145, 148, 174, 178, 191
Geneva, 44
Geodynamics, 189
Geoscience, 40, 42, 147
Geosphere, 41
Glasscrete, 181
Glass waste, 181
Global warming, 4, 36, 40, 49, 101, 107, 111, 166, 226
Goa, 1–11, 16–18, 20, 21, 23–29, 31–34, 40, 52, 59–61, 67–69, 73, 78, 80, 91, 94, 95, 101, 108, 110, 111, 116, 118, 119, 122, 123, 127, 133, 136, 140, 141, 145, 146, 148, 152–156, 160, 166, 172, 173, 176, 177, 182, 183, 185, 191–197, 203, 206–208, 211, 212, 216, 218–220, 223–225
Goa Group, 18, 19
Goethite, 22
Gold, 46, 56, 57, 179
Gondwana, 6, 9
Greenhouse gases, 31, 46, 191
Greenschist facies, 5, 22
Greenstone belts, 19
Greywackes, 18, 20, 21
Groundwater, 36, 46, 48, 52, 59, 73–77, 79–85, 87, 88, 92–98, 100–102, 105–112, 117, 118, 120, 124, 132–143, 153, 159, 160, 175, 177, 197, 201, 223

H

Hazard, 42, 49, 62, 64–67, 104, 175
Hazardous waste, 182, 183, 218
Hematite, 23, 34
High-level bauxite, 30
Hornblendic Division, 5
Humans, 42, 44, 46, 48, 51, 63, 102, 128, 130, 131, 134, 136, 147, 149, 154, 155, 157, 189, 215
Human trafficking, 209
Hydrological cycle, 41, 102

I

Immigrants, 184, 208, 209
Industrial estates, 31, 134–136, 182, 183
Industrial waste, 67, 153, 179, 183
Iron, 5, 8, 21, 23, 24, 26–29, 32, 46, 47, 52, 56–59, 61, 94, 106, 122, 124, 133, 140, 152, 155, 156, 174, 178, 179
Islands, 192

K

Keri, 9, 10, 31, 136, 217
Khazan (*Cantor*), 120, 121, 124, 201, 203, 211, 218
Khondalites, 34
Komatiite, 22

Konkan, 1, 6
Konkan railway, 11, 63, 157
Kundaim, 135, 176, 183, 218

L
Lagoon, 10, 11, 198
Lakes, 10, 74, 81, 97, 116, 122, 127, 141, 202, 209, 212, 218, 221
Landfills, 152, 173, 174, 176, 177, 181, 183, 223
Landforms, 6, 10, 41, 62, 64, 100, 116
Laterites, 5, 6, 29–31, 33, 68, 81–83, 94, 95, 157, 201, 208
Lateritoid deposits, 5
Layered intrusion, 22
Layered mafic-ultramafic complex, 19
Leachates, 134, 173, 174
Lifestyle, 131, 193, 200, 208–210, 214
Lime, 35, 36, 57, 157, 165, 212
Limonite, 22
Lineaments, 16, 29, 100
Lithosphere, 41, 44, 49
Low-level bauxite, 30, 31

M
Madhei, 219
Mafic rocks, 19, 22, 157
Magnetite, 5, 20, 25, 26, 34, 52, 82
Mandovi, 7–9, 24, 33, 34, 79, 81, 94, 105, 118, 120–122, 124, 125, 140, 201, 203, 219
Manganese, 5, 8, 18, 23, 24, 27–29, 56, 57, 59, 61, 140, 149, 179
Manganism, 149
Mangroves, 120, 122, 124, 159, 161–165, 201, 203, 204
Margao, 2, 93, 104, 106, 126, 134, 138, 139, 146, 158, 173, 175, 207
Marshes, 121, 122
Medical geology, 147
Meretrix casta, 126, 159
Metadolerites, 20, 22, 23
Metakaolin, 178, 179
Metamorphism, 5, 22, 26
Microorganisms, 140
Microplastics, 110, 128–131
Migmatitic gneisses, 16
Migrant influx, 212
Mine, 111, 121
Mineral Policy, 51, 53, 56, 59
Minerals, 2, 4, 5, 8, 16, 23, 26, 27, 32–36, 40, 46–48, 50–58, 61, 62, 68, 77, 82, 96, 116, 121, 122, 124, 132–134, 136, 143, 148, 152–155, 157, 159, 162–164, 174, 178, 214, 222
Mining, 4, 23, 27, 29, 31, 34, 36, 40, 49, 51, 53–62, 64, 68, 69, 94, 95, 104, 108, 110, 111, 116–122, 124, 126, 140, 143–161, 165, 179, 182, 200, 212, 214, 221
Mollem Palaeoplain, 105
Mopa-Verna palaeoplain, 30, 31
Morji, 10
Mormugão, 2, 9, 31, 33, 85, 124, 128, 135–138, 147, 173, 175, 185
Moyar-Bhavani shear belt, 16
Mumbai, 7, 63, 195, 215

N
National park, 12, 118
Netorlim, 8, 22, 24, 28
Netorlim-Verlem, 28
Non-biodegradable waste, 128
Non-charnockitic terrain, 16

O
Occupation patterns, 211
Ocean acidification, 42
Older greenstones, 18
Open cast mining, 40, 60, 68, 118, 153
Orlim, 10
Orogenic belt, 7
Overburden, 26, 36, 52, 55, 59–62, 67–69, 118, 120, 144, 154, 156, 162, 165, 179
Oxide facies, 24
Oxisols, 32

P
Palaeoplain, 30, 81
Palghat-Cauvery shear belt, 16
Palolem, 10, 22, 23, 136, 208
Panaji, 2, 22, 23, 29, 93, 104, 106, 121, 126, 134, 138, 139, 146, 148, 158, 166, 173, 175, 203, 204, 207, 221
Pathogenic organisms, 133, 141
Pavers, 177
Peer pressures, 214
Peninsular Gneiss, 17, 18
Peridotite-gabbro complex, 22
Periodic, 100, 163, 165, 224
Pernem, 9, 11, 22, 31, 85, 92, 96, 136, 137, 180, 201, 208

Physical environment, 40, 116, 155, 176
Pilerne, 135, 183
Pisciculture, 95, 121, 126, 153, 154, 163, 201, 211
Pissurlem, 23, 94
Planar surfaces, 30
Planktons, 128
Plastic waste, 128, 129, 131, 132, 180, 181
Plateau, 6–9, 30, 31, 33, 78, 135, 217, 220
Polem, 31, 166
Pollution, 36, 40, 42, 49, 52, 55, 56, 68, 93, 102, 107, 109, 110, 118, 121, 125, 126, 128, 132–134, 136, 140, 141, 143, 145, 147, 150, 152, 162, 165, 173, 174, 176, 180, 182–184, 197, 200, 201, 205, 210, 216, 221, 223
Polymer concrete, 177, 178
Ponda Group, 19, 20, 24, 25, 29
Pozzolans, 178, 179
Precambrian, 7, 9
Present, 9, 28, 30, 34, 35, 51, 52, 59, 93, 101, 105, 106, 110, 129, 133, 144, 152, 156, 163, 164, 166, 173, 191, 192, 195, 196, 198, 222, 226
Proterozoic, 16, 20–22
Protoliths, 157
Protores, 24
Proximal deposits, 20
Pyrite, 24

Q

Quepem, 5, 9, 23, 31, 33, 85, 96, 104, 120, 145, 175, 202

R

Rainwater harvesting, 110
Recycling, 5, 31, 47, 106, 131, 132, 155, 172, 177, 179–182, 197, 221–223, 226
Reforestation, 55, 149, 150
Rehabilitation, 55, 153
Rejuvenation, 197, 226
Replenishment, 93, 100, 159–161
Reserved forests, 118, 119, 200
Reservoirs, 10, 11, 42, 43, 66, 74, 108, 122, 130, 140, 143, 159, 212
Resources, 2, 4, 5, 8, 16, 27, 28, 30, 32, 33, 35, 36, 40, 42, 44–53, 56, 59, 60, 68, 73, 74, 78–81, 83, 84, 92, 93, 98, 102–112, 116, 117, 120, 126, 132, 134–139, 141–143, 149, 150, 152, 154–156, 159, 161, 163–165, 172, 176, 177, 179, 182, 184, 188–191, 193–195, 197, 198, 200, 201, 207–210, 214–216, 219–226
Restoration, 55, 56, 154, 165
Reuse, 5, 47, 93, 106, 107, 109, 111, 123, 131, 149, 172, 177, 179, 182, 183, 197, 223
Ribandar, 158
Rio summit, 44
Risk, 49, 50, 54, 63–69, 130, 197, 216, 221
Rivona, 24, 28
Road traffic, 206
Rocky promontories, 10, 34

S

Sada, 10, 31, 173
Sahyadri, 2, 6, 11
Sahyadri mountains, 2, 6, 11
Sal, 7, 9, 34, 79, 126, 140, 201, 203
Salgini, 28, 29
Sand, 33–35, 64, 77, 82, 83, 126, 139, 157–161, 164, 176, 177, 181, 201, 207, 208, 212
Sand dunes, 203
Sandstone dykes, 64, 65
Sanguem, 7, 11, 25, 33, 85, 94, 96, 105, 120, 146
Sanitation, 5, 105, 106, 111, 142, 217, 221, 222
Sanquelim, 23, 24
Sanvordem Formation, 19, 21
Saprolite, 81, 157
Savordem, 19, 21, 26
Schist belts, 17, 18
Sea shells, 35, 36
Sewage, 5, 68, 105, 106, 121, 125–128, 133, 134, 139, 141–143, 178, 201, 221, 223
Shields, 16, 50
Shimoga-Goa schist belt, 18
Siderite, 22
Silicate facies, 24
Silicosis, 148
Sinistral, 9
Sirigao, 24
Slope, 6, 7, 10, 33, 49, 50, 62–68, 73, 75, 76, 100, 150
Socio-cultural conflicts, 208
Socio-cultural environment, 16, 188
Solid waste, 5, 128, 152, 171, 173–175, 179–181, 205, 217, 225
Sonshi, 26

Southern Granulite Terrain, 16
Spalling, 177
Special Economic Zone (SEZ), 220
Springs, 74, 92, 94, 95, 120, 134, 135
Submergence, 8, 166
Sulphide, 11, 21, 24, 124
Sulphide facies, 24
Supracrustals, 17–19, 21
Suspended particulate matter (SPM), 122, 144, 145, 150
Sustainability, 4, 16, 48, 72, 87, 93, 105, 108, 112, 122, 137, 138, 142, 152, 172, 189–193, 195–197, 214, 215, 222, 223
Swamps, 35, 121, 198
System, 6, 8, 9, 16, 26, 41–43, 48–50, 74, 76, 101, 105–108, 111, 112, 116, 123, 129, 141–143, 148, 153, 155, 160, 162, 176, 180, 183, 190, 191, 194, 197, 201, 206, 221–223, 225

T
Tableland, 30, 31, 33, 81, 82, 135, 183
Talpona, 7, 79, 81
Tamdi Surla, 22
Tanks, 10, 11, 91, 109, 110, 122, 127, 128, 134, 139, 142, 143, 178, 222
Third world, 45, 59, 152, 188, 189
Tholeiitic, 22–24
Threat, 36, 49, 119, 120, 122–126, 128, 136, 142, 155, 158, 159, 161, 166, 172, 180, 209
Tidal creeks, 8
Tidal flats, 120, 121, 211
Tiracol, 7, 34, 79, 81, 158
Tonalite-Trondhjemite-Granodiorite (TTG) gneisses, 16, 20
Tourism, 4, 5, 10, 16, 36, 40, 72, 102, 105, 112, 117, 118, 136, 166, 173, 180, 184, 188–201, 209–211, 213–215, 219, 221–225
Toxic constituents, 124
Toxicity, 130, 135, 153, 173
Toxic metals, 135, 144, 148, 174
Transgression, 9
Transport, 8, 67, 68, 92, 124, 125, 129–131, 142, 144, 145, 147, 150, 153, 156, 177, 200, 205, 206, 210, 213, 214, 226
Trellis, 8
Tuberculosis, 148
Tuem, 36, 120, 121, 164
Turbidite deposits, 18, 20
Tyre-Rubber waste, 182

U
Ultramafic, 19, 22
Usgaon, 145, 146

V
Vageri Formation, 20, 21
Valpoi, 8, 22, 24, 26
Vegetation, 34, 50, 59, 68, 74, 79, 104, 118, 121, 122, 124, 149, 153, 160, 163, 165
Velguem, 25
Verlem, 28
Viliena, 25
Viruses, 133, 141

W
Wasting assets, 47, 51, 223
Water, 2, 5, 6, 8, 10, 11, 18–21, 24, 28, 31, 35, 36, 41–44, 46, 48, 50, 52, 54, 59, 63–69, 72–84, 86–98, 100–112, 118, 120, 122–124, 127–143, 147, 149, 150, 153–157, 159, 161–163, 165, 172–174, 178, 180, 181, 183–185, 193, 195, 197, 202, 206, 208, 209, 211, 212, 215, 216, 219–224
Water conservation, 93, 104, 107, 109, 137
Water quality, 67, 102, 105, 106, 109, 110, 134, 140, 184, 221
Waters, 73
Watertable, 74–77, 82, 84, 86, 88, 94, 95, 97, 100, 108, 120, 133, 139, 159
West Coast Basin, 1, 6, 7, 9, 78
Western Dharwar Craton, 16
Western Ghats Escarpment, 6
Wet lands, 122
Wildlife sanctuaries, 11

Z
Zuari, 7–9, 33, 34, 79, 81, 105, 120–122, 124–126, 140, 184, 201, 217, 219

The manufacturer's authorised representative in the EU is Springer Nature Customer Service Centre GmbH, Europaplatz 3, 69115 Heidelberg, Germany. If you have any concerns regarding our products, please contact ProductSafety@springernature.com

Printed and bound by CPI Group (UK) Ltd, Croydon, CR0 4YY
01/10/2026
02225900-0004